Genetically Modified Forests

FOREST HISTORY SOCIETY ISSUES SERIES

The Forest History Society was founded in 1946. Since that time, the Society, through its research, reference, and publications programs, has advanced forest and conservation history scholarship. At the same time, it has translated that scholarship into formats useful for people with policy and management responsibilities. For seven decades the Society has worked to demonstrate history's significant utility.

The Forest History Society's Issues Series is one of the Society's most explicit contributions to history's utility. The Society selects issues of importance today that also have significant historical dimensions. Then we invite authors of demonstrated knowledge to examine an issue and synthesize its substantial literature, while keeping the general reader in mind.

The final and most important step is making these authoritative overviews available. Toward that end, an initial distribution is made to people with education, management, policy, or legislative responsibilities who will benefit from a deepened understanding of how a particular issue began and evolved. The books are commonly used in education programs throughout the North America and beyond.

The Issues Series—like its Forest History Society sponsor—is nonadvocatory and aims to present a balanced rendition of often contentious issues.

Other Issues Series titles available from the Forest History Society:

American Forests: A History of Resiliency and Recovery
Newsprint: Canadian Supply and American Demand
Forest Pharmacy: Medicinal Plants in American Forests
American Fires: Management on Wildlands and Forests
Forest Sustainability: The History, the Challenge, the Promise
Canada's Forests: A History

FOREST
HISTORY
SOCIETY
ISSUES
SERIES

Genetically Modified Forests

From Stone Age to Modern Biotechnology

Rowland D. Burdon and William J. Libby

The Forest History Society is a nonprofit educational and research institution dedicated to the advancement of historical understanding of human interaction with the forest environment. It was established in 1946. Interpretations and conclusions in FHS publications are those of the authors; the institution takes responsibility for the selection of topics, competency of the authors, and their freedom of inquiry.

Development of this book was co-sponsored with the Institute of Forest Biotechnology, Raleigh, NC. It is published with support from the National Institute of Environmental Health Services, Natural Resources Canada-Canadian Forest Service/Ressources Naturelles du Canada-Service Canadien des forêts, International Paper Company, Plum Creek Foundation, MeadWestvaco Corporation, Arborgen LLC, the Weyerhaeuser Company Foundation, and the Lynn W. Day Endowment for Forest History Publications.

Printed in the United States of America

Forest History Society
701 William Vickers Avenue
Durham, North Carolina 27701
(919) 682-9319
www.foresthistory.org

First edition

Design by Zubigraphics, Inc.

Cover: A young loblolly pine produced by somatic embryogenesis technology.
Photo courtesy of CellFor Inc.

Library of Congress Cataloging-in-Publication Data

Burdon, Rowland D.
Genetically modified forests : from Stone Age to modern biotechnology / Rowland D. Burdon and William J. Libby. -- 1st ed.
p. cm. -- (Forest History Society issues series)
ISBN-13: 978-0-89030-068-8 (pbk. : alk. paper)
ISBN-10: 0-89030-068-2
1. Forestry biotechnology. 2. Trees--Breeding. 3. Forest genetics. I. Libby, W. J. II. Title. III. Series.
SD387.B55B87 2006
634.9'56--dc22

2006030416

Contents

Figures

Sidebars

Foreword

Genetically Modified Forests: From Stone Age to Modern Biotechnology is a story waiting to be told. As the title implies, the story ranges from the time when humans were dependent upon hunting and gathering from the wilds to the sterile laboratory where particular genes are excised from one tree and surgically implanted into another. The result is a tree that has one or more of the characteristics desired by forest geneticists, such as fast growth, desired wood properties, nutritious fruit, resistance to insects and diseases, tolerance to extreme climates and soils, even the ability to extract noxious chemicals from the air and soil.

Numerous books and pamphlets address the domestication of forest trees, but most of those publications cover only one segment of the story, or they are so steeped in scientific complexity that only the graduate student and fellow professional read more than the introductory chapter. The intent of this booklet is to give a brief overview of forest tree improvement, from beginning to today, in such a way that it engages science teachers, students, state and federal policy makers, forestland managers, landowners, and environmental advocates. Regardless of the audience, the intent is education. This publication enables both protagonists and antagonists—not to mention those who haven't made up their minds—to gain familiarity with the often-contentious subject of biotechnology.

The story was melded together by two eminent scientists, both of whom have the ability to communicate to academy-level scientists as well as to lay readers. They span two continents. Dr. Rowland Burdon, from New Zealand, is a quantitative geneticist who has dealt in mathematical and modeling approaches. Dr. William Libby, from the United States, is a forest geneticist. That combination has given us a story that is both engaging and factual. Their booklet serves as a basic text for all nonspecialists who care about our forests and the place of genetic engineering in modern life.

Bob Kellison
President, Institute of Forest Biotechnology
Professor Emeritus, North Carolina State University
Raleigh, North Carolina

Bruce Zobel
Senior Consultant, Zobel Forestry Associates
Professor Emeritus, North Carolina State University
Raleigh, North Carolina

Overview

- Humans have caused genetic modifications in trees by their presence and activities since earliest times. Such influences have been unintentional as well as deliberate. Unintended genetic modifications are much more common than the relatively few purposeful modifications thus far accomplished by breeding and modern biotechnology.

- The genetics of many tree species have been purposefully modified, beginning more than 5,000 years ago with the selection and propagation of individuals that produced higher yields of fruits, nuts, and oils. Trees have also been genetically altered to offer aesthetic values.

- In the 20th century, purposeful genetic modification of forest tree species for timber production was achieved largely by conventional breeding.

- In the 21st century, biotechnology—defined as anything that combines technology and biology—is aiding our understanding of plant genetics, improving our understanding of evolutionary relationships among species, and allowing us to preserve the genetic material (DNA) of rare trees, all without causing any genetic modification.

- Genetic engineering, a form of biotechnology that extends traditional tree breeding efforts, is making it possible to achieve genetic changes by inserting small fragments of DNA into the DNA of a recipient clone.

- Forest trees that are intentionally genetically modified—whether by conventional breeding and/or through biotechnology—will generally be planted in intensively managed plantations.

- Wood from tree plantations can substitute for alternative materials that, in their extraction and manufacture, require more energy inputs and release more toxic materials and carbon dioxide.

- Demand for forest products, such as timber, paper, and firewood, continues to grow. This demand places considerable pressure on the world's forests. Growing wood products in plantations makes it possible to preserve other forests, such as old-growth and high-biodiversity forests, and still meet human needs for wood.

- Genetically modifying trees for intensive timber production presents the opportunity to better meet future human needs for wood while allowing more natural forest area to be set aside in reserves and parks.

- The environmental and technical risks of growing genetically engineered trees raise significant concerns.

- Engineering sexual sterility into genetically modified trees is one option to help prevent their engineered genes from "escaping" and altering natural ecosystems; escape can also be controlled by isolating such trees, harvesting them before they can reproduce, or both.

- Engineering sterility may also serve to enhance wood production.

- Regulatory systems to help manage the risks of biotechnology and address public concerns are appropriate.

Introduction

This booklet presents a historical account of human-caused genetic modifications of forest trees. In the second half of the 20th century, trees were "improved" through conventional breeding techniques, in much the same way livestock are bred to bring out certain desirable characteristics in their offspring. As science and technology have advanced, so have the pace and extent of genetic modification. A current and growing issue in forestry is the application of modern biotechnology, including genetic engineering, to forest trees.

The term *biotechnology* came into common usage in the 1980s, and its several definitions continue to change. Broadly defined, it is anything that combines biology and technology. Biotechnology has become more controversial as the level of the technology has increased. For example, cloning fruit trees by grafting is low-tech biotechnology, long practiced and not particularly controversial. But transferring individual genes from one species to another is high-tech, very recent, and more controversial.

Biotechnology has the potential to address a major problem—the predicted insufficiency of today's forests to meet the wood needs and desires of tomorrow's human population. Earth's population has doubled twice in the 20th century, now exceeds 6 billion people, and is projected to exceed 9 billion around the year 2050. Those 9 billion to 10 billion people are likely to require an enormous amount of lumber, paper, and other wood products—particularly if they increase their consumption to the level of Americans, who use about four times as much wood per person compared with the global average.

It is often said that technology will find replacements for wood. The current replacements are various metals, brick, cement, plastics, and alternative fibers. Compared with wood, virtually all of these replacements require more energy, emit more greenhouse gases, and introduce more toxic materials into the environment during their extraction and manufacture. If new sources of energy are abundant and inexpensive, and if inexhaustible sources of alternative raw materials are found, then per-person needs for wood can be reduced through substitution—almost certainly with the unintended consequence of a warmer, more toxic world. But if, as seems more likely, energy becomes more expensive and nonrenewable raw materials are less available, wood could increasingly substitute for those other products, per-person use of wood would be even higher, and Earth would probably be cooler and less toxic.

Growing more wood, then, may help meet the needs of our expanding population with less damage to the environment. Intensive plantation forestry—in effect, growing trees like corn or soybeans—is an economical way of producing large quantities of high-quality wood. But it can also serve wider conservation values: an abundant supply of wood from tree plantations should relieve pressure on natural forests that have important biodiversity, ecological, aesthetic, and other values. As well, the plantations themselves can and do provide important ecological services. The existence of intensively managed plantations is the prerequisite for any intensive genetic improvement. Conversely, genetic improvement can make intensive plantation forestry worthwhile.

Few would question the use of genetic engineering if it is used to protect the environment and improve living standards, but there are proper concerns about unintended consequences. The transfer of genes between unrelated species holds tremendous promise; it also presents risks that make biotechnology one of the most controversial issues in forestry. This booklet reviews the issues surrounding plantation forestry and the genetic modification of forest trees.

Terms that may be unfamiliar to readers are highlighted in **boldface** type on their first significant appearance in the text and defined in the Glossary.

Chapter 1

Historical Efforts to Improve Trees

In the millennia when hunter-gatherer humans lived lightly on the land, their impacts on forests were little different from those of other forest creatures. After they began using wood for cooking, heating, shelter, fortification, and transportation, their choices in the trees harvested increasingly influenced both the species composition of the forests and the genetic constitutions of selected species. Early humans also modified vegetation by setting fires, and much later, some societies began to purposefully manage their forests by **selection** and **interplanting**. Some even established forests where forests did not naturally exist.

Human influence on the forest can be indirect and unintentional as well as deliberate. When we harvest the more valuable trees, we leave those with defects and deformities as the parents of future generations, and over time, such genetic modifications can degrade the forest. Genetic changes in tree populations also occur in response to such things as air pollution and climate change, even in remote reserves.

Planting trees for fruit and to make gardens and sacred groves began 6,000 or more years ago, but planting trees to replace or create forests was exceptional. As people ran short of wood, a common response was to acquire the wood elsewhere, often by conquest. This imperialistic response led Babylon's King Hammurabi to invade neighboring kingdoms in Mesopotamia 4,000 years ago. Perhaps the first recorded instance of reforestation to alleviate dwindling wood supplies dates to 2,300 years ago, when the Ptolemaic Dynasty in Egypt established government nurseries and sponsored massive tree-planting programs. Since that time, **clones** of easily rooted poplar and willow have been planted in Asia, the Middle East, and around the Mediterranean for erosion control, basket materials, and fodder and bedding for animals.

Starting in the 1200s, Chinese-fir was planted in several regions of China for high-value timber uses. A redwood relative, this tree has long been one of China's most valuable tree species. Like the coast redwood of North America, it sprouts from the stump. Cuttings from Chinese-fir stump sprouts readily root when inserted into moist soil in appropriate conditions. About 800 years ago, farmers in southeastern China took note of the better Chinese-fir trees on their land. After those trees were harvested, the tops of the vigorous sprouts that originated from the stumps could be planted as unrooted cuttings, and their

Figure 1A. *Modern almonds. Almonds are among California's most valuable orchard crops. Their wild ancestors, found in eastern Mediterranean forests 10,000 years ago, had small, bitter nuts that were deadly poisonous. Almonds have been purposefully genetically modified perhaps longer than any other tree species. Photo by Iris Libby.*

already-rooted basal portions could be detached and planted. Using both types of regeneration, farmers would thus establish new groves from their best trees. Locally adapted clones were used in 16 Chinese provinces with greatly different environments. This is the first recorded example of purposeful forest tree improvement.

In 1664, John Evelyn published *Sylva*, a work commissioned by Britain's Royal Society to assess deforestation and explore reforestation as a means of ensuring a sufficient supply of timbers for the Royal Navy. He wrote, "Chuse your seed…from the tops and summities of the fairest and soundest trees." This was an early statement suggesting that foresters should at least maintain quality equal to that of the parents and even improve upon the average quality found in the forest. Evelyn knew that the larger, healthier seeds develop near the tops of trees. He may have also been aware of the deleterious effects of **inbreeding** and surmised that seeds from the upper crowns of wind-pollinated trees were less likely to be inbred. By the 1700s, following his advice, the English were planting and husbanding oaks for ship timbers.

In 1831, nearly two centuries after the publication of *Sylva*, Patrick Matthew, a forester with much experience in British **plantation** forestry, presented the principles of natural selection and described how populations adapt and differentiate in response to contrasting environments. *On Naval Timber and Arboriculture* appeared 11 months before Charles Darwin sailed on the voyage that led him to consider and understand evidence for **evolution**. In later correspondence with Matthew, Darwin acknowledged that Matthew had presented most of the important ideas that he, independently but 28 years later, set forth in *On the Origin of Species*.

Also in the 1800s, German landholders established large plantations of conifers to offset regional depletion of native forests. In 1879, Max Kienitz urged German foresters to improve their native species rather than attempt forest improvement by planting nonnative, or **exotic**, species. Despite Kienitz's admonition, European foresters recognized that, for reasons still not fully understood, conifers from western North America grew faster, better, and larger than did their native conifers. In the late 19th and early 20th centuries, exotics like Sitka spruce and Douglas-fir, both native to western North America, were widely planted in Western Europe. In contrast, large plantings of Norway spruce, while native to Europe, were often incorrectly sited and performed disappointingly in Germany and parts of Switzerland.

Figure 1B. *Traditional clonal planting of Chinese-fir. This highly productive plantation in Nanping, Fujian Province, was established in 1919 with rooted and unrooted stump sprouts of long-selected clones and by 1996 had become a reserved stand. Professor Minghe Li provides scale. Photo courtesy of Minghe Li, Wuhan University.*

In southwestern France in the 19th century, 2 million acres of maritime pine plantations were established both to create a resin industry and to stabilize coastal dunes and keep sand from encroaching on agricultural lands. In the early 1900s, several countries in the Southern Hemisphere—most notably South Africa, Australia, New Zealand, and Chile—began establishing extensive plantations of nonnative tree species to counter depletion of their native forests and, eventually, to build export industries.

EARLY FAILURES

During the early centuries of planting trees in forests and plantations, wild seeds were generally collected near where they would be used. As communication and transportation improved in the 19th and 20th centuries, however, seeds were often collected at greater distances from their intended planting sites. This practice frequently proved to be a serious mistake, since the seedlings from distant populations sometimes encountered combinations of climate, soils, insect **pests**, and **pathogens** to which they were not adapted. That result sparked interest in matching seeds of the desired species to the site.

Provenance is the French word for "origin" or "source"—in our context, the location of the native tree population(s). The seeds for a particular tree plantation may come from a source that is not the provenance. For example, Scots pine is a nonnative tree that grows well in upstate New York; if Scots pine seeds are collected from a plantation near Albany, New York, the source would be Albany but the provenance—the original home—might be Latvia, in north-central Europe.

At first unaware that provenance was critical to the success of plantation forestry, foresters throughout the world often blamed plantation failure on drought. This "great excuse," as Bruce Zobel and Jerry Sprague call it in their insightful history, *A Forestry Revolution*, was often invoked even though seedlings from the wrong provenance had been planted; compounding foresters' problems, the seedlings were often in poor condition when planted, site preparation was inadequate, the planting was done badly, or competing species were not controlled.

Unlike fanciers of ornamental flowers and farmers competing for blue ribbons at the state fair, many early-20th-century foresters resisted the idea that genetic variation might be important, and some denied it even existed. Even after the importance of provenance was recognized, many foresters had little appreciation of heritable variation. Much of the problem can be traced to forestry education: in most university curricula, genetics was not a required or even recommended course until the 1960s or 1970s.

Figure 1C. *Monterey pine cones. Typical opened (upper row) and closed cones were sampled from the five native populations of Monterey pine. They had been grown in the same place to remove major environmental differences. The populations are, from left, mainland Año Nuevo, mainland Cambria, Cedros Island, Guadalupe Island, and mainland Monterey. Fossil cones deposited on the mainland before the arrival of Native Americans look very much like the present cones from the two islands, being small and often symmetric. The cones on the mainland have since evolved to become larger and more asymmetric and have thickened outer scales, all of which protect the seeds from the heat of fires. This seems likely to be an adaptation by the mainland populations to the frequent fires that Native Americans used for 8,000 or more years to facilitate hunting and improve berry crops in the understory. Photo by Iris Libby.*

Figure 1D. *Elevational transect provenance test. As one of its first projects, the Institute of Forest Genetics (see page 19) collected seeds from nearby native ponderosa pines at intervals of 500-foot elevation along a transect in California's Sierra Nevada, from above Sacramento to near Lake Tahoe. Replicate trials were planted in 1938 at low, medium, and high elevations. In 1984, Hans Roulund, from the National Arboretum at Hørsholm, Denmark, stands in the middle-elevation trial between a plot of local mid-elevation trees (right) growing well, and higher-elevation trees (left) growing less well. Relative performance is reversed at the higher-elevation replication. This experiment was a catalyst in California, resulting in seed-zone maps and seed-transfer guidelines. Photo by Bill Libby.*

Figure 1E. *Range-wide provenance test of an exotic pine. This 10-year-old provenance test of contorta pine was planted in 1958 on the Kaingaroa Forest in central North Island, New Zealand. Contorta pine has a large native range, from sea level to above 7,000 feet elevation in California, and from California north to Alaska and inland to the Rocky Mountains. Large differences in survival and growth among blocks from different provenances are evident in this trial. The planting site is reasonably uniform, as shown by the border row of trees, all from the same provenance. Photo by H.G. Hemming, New Zealand Forest Service.*

Once trees were being planted for commercial purposes, however, investment in planting productive, highly valued species made economic sense: such trees would increase financial returns. Improved genetic stock and more intensive management thus go hand in hand, and the second half of the 20th century would see refinements in both the methods of plantation forestry and the techniques of tree improvement. The resulting increase in timber production from intensively managed plantations provides one possibility for slowing destruction of the world's natural forests.

Chapter 2

The Science of Tree Breeding: A Primer

Breeding for "tree improvement" means genetically modifying a population of trees to better serve human purposes. Trees with straight trunks and small branches, for example, are easily processed and produce a high proportion of lumber, so tree form has been one target of tree breeders. Improved seedlings are planted in prepared sites and perhaps fertilized, pruned, thinned, and otherwise tended in ways that promote their growth and stem quality.

Three terms have similar meanings but are distinguished as follows in this booklet:

Genetic modification is any process that changes the genetic constitution of populations or individuals; it implies influences and processes associated with or applied by humans. It can include unintentional genetic changes, both beneficial and detrimental, as well as purposeful activities meant to genetically change populations, **breeds**, or individual clones.

Genetic improvement is a purposeful genetic process meant to produce **characteristics** desired by humans. For forest trees, it has, until recently, been accomplished by identifying appropriate populations, doing selective breeding, and deploying the resultant offspring in tests and plantations.

Genetic engineering can refer to all purposeful genetic improvement activities, but as used today, the term involves inserting specific short fragments of genetic material into the otherwise intact genetic constitution of an organism—in our case, forest trees.

Genetic engineering is addressed in Chapters 5 and 6, but the conventional approach to tree improvement, which accelerated with the appreciation of Mendelian principles of heredity, is the logical first step for discussion.

SEXUAL REPRODUCTION IN PLANTS

In the 19th century, Gregor Mendel, an Austrian monk and amateur botanist, studied garden peas, but the principles of reproduction and genetics that he discovered for that plant are the same as for other multicellular organisms, including trees. However, unlike the pea plant, which normally self-fertilizes because the male and female organs are within a flower that remains closed until pollination is complete, most important temperate and boreal tree species are **monoecious**. This means that male and female organs are separate from each other on the

Figure 2A. *Male and female flowers of a tree. The female strobili (flowers) of loblolly pine (left), found in the upper reaches of the crown, are at the optimum stage to receive pollen. Male strobili (catkins) from the same tree are found in the lower branches (right); those shown here are just before pollen release. The separation of female and male parts within a tree's crown reduces the probability of selfing, the most severe form of inbreeding. Controlled pollination occurs when pollen from either a single tree or a known mixture of several trees is applied to female strobili that have been bagged to exclude unwanted pollen. Photo courtesy of NCSU-Industry Cooperative Tree Improvement Program.*

same tree, and thus fertilization of the females by pollen from other trees is common. It is still possible, though, for pollen to reach the receptive female organs on the same tree, producing highly inbred offspring called **selfs**.

Among humans, **inbreeding** has long been recognized to have unfortunate biological and sociological consequences—hence incest taboos—and the biological principle holds for trees as well. Abundant research has shown that with few exceptions, inbred trees do not grow as well and have poorer health than **outcrossed** trees—that is, trees coming from unrelated parents.

Self-fertilization, the most extreme form of inbreeding, is naturally controlled in several ways. The female organs most often occupy the upper portions of the tree's crown, and their male counterparts are found below; wind-borne pollen is less likely to drift upward than horizontally. In nature, selfing is further controlled by—singly or in combination—sexual incompatibility, failure of inbred seed to develop and germinate, or failure of the inbred seedling to grow normally even if the seed does germinate. Some tree species are **dioecious**, meaning that male and female flowers form on different trees, which precludes self-pollination.

The Linnean classification system for plants, and especially forest trees, is based largely on flower form. Trees are classified as either **gymnosperms** or **angiosperms**. The pines, spruces, firs, and cedars are gymnosperms, characterized by having naked **ovules** in structures called **strobili**. Oaks, birches, apples, maples, and most

Figure 2B–C. *Inbred and outcrossed redwoods. Following the observation of large differences among adjacent redwood clones, controlled pollinations were conducted to produce outcrossed and inbred redwoods with parents in common. A "self" is 50 percent inbred, produced when pollen from a tree fertilizes the female part of that same tree. In Figure 2B (above), parent R37 is both the mother and the father of this tree, planted in 1982. Parent R17 is from the same provenance but unrelated to R37; their mating thus produced an outcross (Figure 2C, below). The photographs were taken after 20 years of growth in the field. The identifying tags, "R37 self" and "R17 × R37," are at the same height on these adjacent trees, both tags being 5 inches square. Although some selfs in this trial grew better than the tree in Figure 2B, others were worse, and without exception all selfs were smaller than their paired related outcross. Furthermore, many of the selfs but few of the outcrosses had died in the first 20 years. Photos by Bill Libby.*

other broad-leafed trees are angiosperms; they have enclosed ovules within true flowers. Some angiosperms have **perfect flowers**—both male (pollen-bearing) and female (seed-producing) organs are present in the same flower—and are almost always pollinated by insects or, in some situations, by bats. Yellow-poplar is an example of an angiosperm with perfect flowers. Angiosperms with imperfect flowers—that is, flowers that produce either pollen or seed—are usually wind-pollinated, like the gymnosperms. Oak trees, for example, have imperfect flowers but are monoecious.

Pea plants have perfect flowers. Beginning in 1843, through carefully controlled pollination of peas over a number of generations, Gregor Mendel worked out and in 1865 published the rules of single-**locus** inheritance—that is, for **traits** that are governed primarily by a single **gene** at one location on a **chromosome**. For the next 35 years, his work was overlooked by most and not understood by others. By 1900, however, science had sufficiently advanced that Mendel's rules

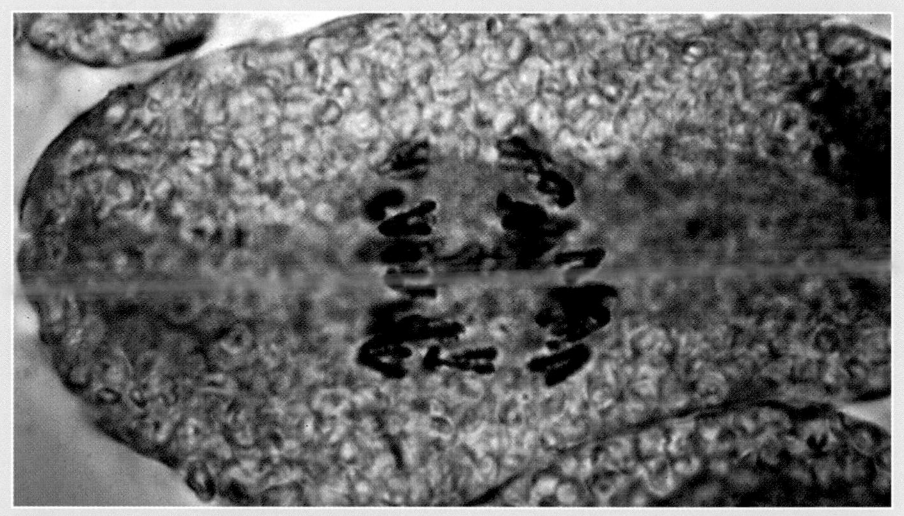

Figure 2D. *Nuclear division in Douglas-fir. As with all plants and animals, most tree genes are part of structures within the nucleus of each cell called chromosomes. Douglas-fir normally has 26 chromosomes. The chromosomes are composed of proteins and deoxyribonucleic acid (DNA). The genes are discrete sections of the DNA. Shortly before cell division, the chromosomes and their contained DNA duplicate, creating pairs of chromosomes. Then, one chromosome of each pair is pulled toward one end of the cell, the other chromosome of each such pair is pulled toward the other end, and the cell divides between these two complete sets of chromosomes. Each new cell thus has the same chromosomes and genes. Photo courtesy of Phil Haddock.*

suddenly made sense to many biologists, and the field of genetics was formally established in the first decade of the 20th century. During the next two decades, the importance of single genes and the mechanism of inheritance dominated biological thought.

EVOLUTIONARY PROCESSES

In the 1920s and early 1930s, Sewall Wright in the United States, R.A. Fisher and J.B.S. Haldane in England, and S.S. Chetverikov in Russia—each working independently—developed a synthesis of Mendel's rules concerning discrete variation of traits governed by one or a few genes, and the rules that better apply to continuous variation governed by genes at many loci. Such continuous variation is typical of many biologically and economically important traits, including growth and yield in most crop plants and forest trees.

Chetverikov, a naturalist, summed up in a 1927 essay how genes work and evolve in populations: mutations originate and replenish the large amounts of genetic variation typical of most natural populations, and discrete single-locus inheritance can produce and maintain continuous variation by genes at many loci acting cumulatively and interacting with each other. The concept that frequencies of **alleles**—different forms of the same gene—help distinguish the genetic constitution of populations was born.

There are four major processes that can change allele frequencies: mutation, migration, selection, and genetic drift.

Mutation is the source of all genetic variation, and as such it is the raw material for evolution. Mutations of a gene can occur in response to internal or external influences and change some element of the organism's genetic constitution. Over only a few generations in sexually reproducing organisms, such as trees, mutation is a minor force. Its effect is usually to increase genetic variation within a population, often by introducing a new allele.

Migration is the movement of genes from one population to another. Genes emigrate *from* and *im*migrate *into* a population. Emigration has no effect on either allele frequency or genetic variation in the donor population, but the effects of immigration can be large. Immigration changes allele frequencies within the recipient population toward those of the population from which the migrant came, and it usually increases genetic variation within the recipient population.

Italian Cypress

Mediterranean cypress is native or naturalized throughout the Mediterranean region, and it typically has a broad crown. The crown form is largely controlled by the two forms of a single gene (alleles: *Cb = broad crown; Cn = narrow crown*), which occur at a single locus in its chromosomes. The normal broad crown develops when a seedling's two genes at that locus are the alleles *CbCb*; a narrow, columnar form develops when the alleles are *CnCn*; and an intermediate-width crown develops when a seedling has one of each allele, *CnCb*.

The Roman Emperor Diocletian was much taken with the narrow columnar form of Mediterranean cypress and had it planted at many sites in his empire. Now common in many parts of the world, it is known as the ornamental "Italian cypress."

During the 17 centuries since Diocletian's rule, pollen from columnar trees has blown into surrounding areas, and the *Cn* allele has become established in many native and naturalized populations of cypress. Trees with intermediate-width crowns occur frequently in forests near his old palaces and other planted sites, and some narrow, columnar trees now occur naturally in those forests.

Figure 2E. *Narrow-crowned Italian cypress* (CnCn) *on the Spannocchia estate in Tuscany, Italy. This form of the Mediterranean cypress has become widespread, thanks to the intervention of humans who preferred its narrow crown for purely aesthetic reasons. Its wind-disseminated pollen deposited on the strobili of wide-crowned Italian cypress* (CbCb) *produces an intermediate crown type* (CbCn). *Photo by Steven Anderson.*

Selection is the process that favors survival and successful reproduction of some individuals over others; it generally decreases genetic variation. Selection can be a powerful force in the short term and usually is a dominant force in the long term. An appreciation of selection was Charles Darwin's best-known contribution to an understanding of evolution.

Genetic drift occurs when only a few individuals reproduce or colonize a new site and create the population's next generation. This phenomenon can drastically change allele frequencies, and it usually reduces genetic variation within a population.

POLITICAL BIOLOGY

At first Chetverikov's thesis was not widely appreciated, in part because it was published in Russian. Then, in Russia, his concepts became politically incorrect. He was dismissed from his university position but nevertheless continued some genetic studies while teaching math in a junior college and consulting with a zoo in the Ural Mountains.

In August 1948, a decree from the Soviet Academy of Sciences, published in *Pravda*, revised the kinds of science to be taught and the kinds of research to be permitted. T.D. Lysenko, a plant physiologist, was put in charge of genetics. In support of the social transformation of the U.S.S.R., heredity was declared to respond to acquired or imposed characteristics—for example, if the parents worked hard, their children would inherit an ability to work hard—and the accepted roles of genes and chromosomes in heredity were officially repudiated. This dogma adversely affected agriculture in the U.S.S.R. for decades. Chetverikov was again dismissed; he became impoverished and lost his eyesight. Just before he died in 1959, the Soviet government relaxed its opposition to genetics, and he received some recognition.

In the meantime, however, Lysenko's politically inspired theories had influenced Chinese forest policy of the 1950s and 1960s. The use of long-selected, well-tried clones was officially discontinued; it was ordered that "more natural" seedlings be used for reforestation. The resulting Chinese-fir plantations suffered insect and disease damage, with reductions in wood harvests. Luckily, many remote farmers ignored orders from authorities in Beijing, and most of the excellent clones are still available and again being used.

Figure 2F. *Defiance of political biology. Near Jianhua, Hunan Province, the local Miao people ignored instructions from Beijing and continued planting traditional Chinese-fir clones rather than the mandated seedlings. This 20-year-old plantation, photographed in the 1980s, has excellent wood and no observed insect or disease damage. Photo courtesy of Minghe Li, Wuhan University.*

PATTERNS OF GENETIC VARIATION

One of Chetverikov's conclusions concerned genetic constitution—all of the genes in an individual or population or species: how those genes are organized and affect each other, which alleles of those genes are present, and in what frequencies.

The genetic constitutions of species are not static; some degree of genetic change is normal. Genetic changes in forest species repeatedly occurred prior to the arrival of humans, and ever since, forest species have adapted to the actions of humans, just as they have adapted to other creatures that lived in, on, or among them.

When considering the genetic modification of forest trees by humans, it is useful to consider the levels of a species' genetic organization that might be affected. Different evolutionary forces and genetic processes affect each of four levels of genetic variability: (1) major geographic populations, (2) local populations, (3) **families**, and (4) individuals within families.

Major geographic populations are, by definition, separated by distance. The environments of widely separated places are usually substantially different in such things as day length and climate. In the absence of humans, genetic material from distant populations is unlikely to have much effect on adaptive traits shaped by natural selection because natural, long-distance dispersal of seeds and viable pollen is dependent on wind and migrating animals and birds. Moreover, pollen or seed that does come from afar is likely to be ill-adapted genetically to the local conditions. For these reasons, mutant alleles that become established in a local population are unlikely to spread rapidly to distant populations.

Local populations are close enough to each other that seeds and pollen can occasionally or even frequently reach other nearby populations and thus effect genetic change. Genetic differences caused by weak local selection are generally overwhelmed by massive migration of genes from the pollen of adjacent populations. But even repeated migration of differently adapted genetic material sometimes causes little change in a population: strong selection—for example, in neighboring populations on two very different kinds of soil—can maintain adaptive genetic differences between them.

Genetic changes due to chance events are particularly important in populations that are small, and also in currently large populations that were established from just the few parents that colonized the site or from only a few survivors of some stand-replacing event. Sometimes, such apparently random local variation produces pleasant surprises. For example, compared with Australia's many native river redgum populations of eucalypts, the Lake Albacutya population

Figure 2G. *A family provenance trial. Norman Borlaug, Nobel Peace Prize laureate for his pioneering of the "green revolution," inspects plots of open-pollinated Douglas-fir families from the same and different stands on the Bitterroot National Forest in Idaho in 1973. Photo by Bill Libby.*

of these trees is small and not particularly impressive, yet it has proven to be broadly adapted and among the very best in provenance tests and plantation programs in several parts of the world where river redgum is now grown.

By the late 1960s, population biologists were increasingly finding the genetic structure of tree species to be of research interest and recognizing the importance of within-population genetic variation. It was becoming clear that long-lived species typically maintain more within-population genetic variation than do short-lived species. Among the longer-lived species, those that stay in one place and endure seasonal and cyclical changes in their local environments typically maintain more within-population genetic variation than those that migrate each year or otherwise modify their environments. Being long-lived organisms that remain in one place, trees were expected to maintain very high levels of within-population genetic variation. With a few interesting exceptions, subsequent research has confirmed that expectation for forest tree species.

Tree breeders have taken that finding and used this high level of genetic variation for the purposes of tree improvement. The work of tree breeders is the subject of the next chapter.

Chapter 3
Conventional Tree Improvement

Purposeful breeding of trees is very much based on plantation forestry. Such breeding began in the second and third decades of the 20th century, in northwestern Europe and the northeastern United States. This early work mostly involved producing, testing, and selecting hybrids—first poplars, then pines.

A **hybrid** is the progeny of parents from genetically different populations, either of the same species **(intraspecific)**, or between different species **(interspecific)**. Hybrids between species sometimes combine desired characteristics from each species. Loblolly pine, for example, is superior to shortleaf pine in growth and utility but is more susceptible to fusiform rust, a fungal disease. A hybrid of the two species is valuable in areas of high fusiform rust incidence even though it typically grows more slowly than its loblolly pine parent. A further refinement to optimize growth and disease resistance can be obtained by **backcrossing**—in this instance, mating the hybrid back to a loblolly pine. The procedure can be continued, each time using the most advanced hybrid with a different loblolly pine parent (to prevent inbreeding), until a population of trees is produced that has the growth performance of loblolly pine but the disease resistance of shortleaf pine.

In the early 20th century, hybrid corn was becoming a success story in crop breeding, and hybrids between poplar species had often shown outstanding vigor. J.G. Eddy, a visionary forest owner from Seattle, hired the famous fruit-tree breeder Luther Burbank to locate an appropriate site for a forest-tree breeding station. Based on favorable soil and climate, they selected Placerville, California, in 1925. Much of the early work at Placerville focused on producing and testing interspecific hybrids of pines. Techniques for controlled pollination were developed and relationships among species of pine were clarified. Very few of those pine hybrids proved useful in plantations. They were expensive to produce and rarely outperformed a parent species in its native range.

In 1935, the site was transferred to the USDA Forest Service and became the Institute of Forest Genetics. The private and federal occupants of this site have an unbroken history from 1925 to the present day of providing staff, support, and facilities for advancing research and development of forest genetics, tree improvement, and forest biotechnology.

Figure 3A. *An early hybrid poplar clone. Brian Barkley, of the Ontario Fast Growing Hardwoods Program, stands in a nine-year-old planting of the "robusta" clone in southeastern Ontario, Canada, in 1983. This hybrid was produced, tested, and selected in Maine in the 1920s. More recent clones are more productive, but robusta is known to have high-density wood and veneer log form and is well suited for agroforestry operations. Photo by Bill Libby.*

Figure 3B. *The Institute of Forest Genetics at Placerville, California, in 1937. These office, laboratory, and residence buildings were constructed after the Eddy Tree Breeding Station was transferred to the U.S. Forest Service. Since 1925, the Institute has provided staff, support, and facilities for research and development advancing forest biotechnology, forest genetics, and tree improvement. Photo courtesy of the U.S. Forest Service Institute of Forest Genetics.*

THE SCANDINAVIAN IDEAS

The foundation for within-species breeding of forest trees was laid in Scandinavia during the 1930s and 1940s. The writings of C. Syrach Larsen, director of the Danish National Arboretum, began to attract attention in the 1930s. With the 1943 publication of his *Genetics in Silviculture* and its 1956 English translation, a much larger readership became familiar with techniques of breeding and propagation of forest trees and with the possibilities offered by applications of genetics in forestry.

Larsen used grafts and rooted cuttings to create clonal "tree shows." The obvious and consistent contrasts between different clones grown side by side clearly demonstrated the important influence that genetics exerts on growth, form, and other important traits of trees. That powerful evidence helped convince both foresters and skeptics that tree similarities and differences are significantly governed by genetics.

In Sweden, a society for practical forest tree breeding was formed in 1941. Åke Gustafsson, an eminent geneticist specializing in agricultural crop breeding,

Figure 3C. *C. Syrach Larsen. In 1964, one of the very earliest pioneers of forest genetics stands next to a "hybrid" of grand fir and white fir that he produced 40 years before. This tree is growing near Charlottenlund, Denmark. This and other evidence lead some to conclude that grand fir and white fir are parts of the same species. Photo by Phil Haddock.*

assumed leadership of the forest genetics program in Stockholm in 1946. Following World War II, foresters of various countries returning to peacetime pursuits became aware of the ongoing forest-tree breeding efforts in neutral Sweden. By about 1950, the Scandinavian protocol for tree breeding had been studied and adopted by many forestry organizations worldwide. Its concepts made intuitive sense, a feature that led to widespread understanding and acceptance.

The Scandinavian protocol consists of four steps:

1. *Find and select* trees of unusually good size, stem form, health, and other desired characteristics from existing populations of the desired species. These are called **plus trees**.

2. *Concentrate and multiply* the genes of those plus trees in **seed orchards**. The selected trees are usually propagated by grafting upper-crown twigs onto rootstocks (one- to four-year-old seedlings). Grafting mature twigs from the tops of the trees brings the reproductive structures close to the ground for human-assisted breeding and seed collection. Cloning each plus tree in many grafted copies greatly increases pollen and seed production from the selected clones. Also, intermixing grafts of the various clones serves to increase outcrossing (and thus decrease inbreeding). The number of different clones in operational seed orchards varies from about 15 to 100 or more, depending on species, region, and the number of seed orchards in a physiographic area. The number of copies of each clone (and thus seed orchard area) is determined by the annual seed-production needs.

3. *Evaluate and rank* the usefulness of the plus-tree clones as parents by growing their offspring together at several sites until they reach a size and age at which important traits can be evaluated. These **progeny tests** serve the primary function of allowing the plus-tree clones to be ranked for continued inclusion in or removal from the seed orchards. They also contribute important information about inheritance of various traits and create some outstanding progeny for future tree-breeding efforts.

Meanwhile, the grafted clones are growing in the seed orchard and beginning to produce abundant seeds.

4. *Produce abundant genetically improved seeds* by removing the lower-ranked clones from the seed orchard, based on the performance of their progeny. After one or more rounds of such refinement, only the top-ranked clones remain to be parents of seeds used for establishing plantations. Not only do the seeds from such an orchard have outstanding progeny-tested plus-tree mothers, but also most or even all of the pollen supplying their fathers' genes comes from the other outstanding plus-tree clones in the seed orchard.

Figure 3D. *Monterey pine NZ55 in 1982. This plus tree imparts to its offspring outstanding growth rate and a wood property that is substantially less energy-demanding in mechanical pulping. It was found in a corporate plantation near Kinleith, New Zealand. Its genes are now recombining in advanced generations of many different breeding lines in several countries. Photo by Bill Libby.*

Figure 3E. *A plus tree in 1971. Flooded gum (*Eucalyptus grandis*) is being grown in plantations in many tropical and subtropical regions. This outstanding native plus tree (left center) was found near Coff's Harbour, New South Wales, Australia. Photo by Bill Libby.*

COOPERATIVE TREE-BREEDING PROGRAMS

As people entered the period of recovery and optimism after World War II, they began to address some looming problems. Among these were a rapidly growing human population in need of housing, other material goods, and food. The recognition that wood was a renewable resource with a multitude of known uses gave impetus to plantation forestry. To increase the quantity and quality of plantations, tree-improvement programs were initiated. The aim was to select high-performing trees, concentrate them in seed orchards, test them, and then use the best trees to produce large quantities of improved seeds.

In 1950, Åke Gustafsson, the Swedish geneticist, gave a catalytic talk in Houston on breeding agronomic crops. One component of the talk was on the application of agronomic crop breeding to forest trees. The governor of Texas, the director of the Texas Forest Service, and several industry leaders were in attendance. They were so impressed that it was decided to establish at Texas A&M University a regional tree-improvement program based on agricultural methodology. The program was jointly funded and supported by some major

Figure 3F. *Seed orchard. Mature seeds of loblolly pine are being collected with the aid of a tree shaker. The vibration of the tree-shaker head sends oscillations up the tree that causes the cones to separate from the branch to which they are attached. The cones are then collected and transported to the extraction plant, where the seeds are extracted, cleaned, and stored. This procedure finds wider application in slash pine than in loblolly pine because the cones are more easily dislodged by vibration. Loblolly pine cones are more commonly collected by tree climbers and operators in bucket trucks, with each cone being individually separated from its place of attachment. Photo courtesy NCSU-Industry Cooperative Tree Improvement Program.*

Figure 3G. *Early field-grafting trials. Bruce J. Zobel and John W. "Jack" Duffield, pioneers in tree improvement and forest genetics, discuss the development of seed orchard technology in 1967 at one of many seed orchards that were being established in the southeastern United States. Photo by Bill Libby.*

forest product companies, the Texas Forest Service, and Texas A&M. In 1951, Bruce Zobel, a young Ph.D. from the University of California-Berkeley, was hired to head the new cooperative.

Zobel developed the model for cooperatives among forestry groups. The university contributed facilities, and its strong science programs provided an environment and courses appropriate for the graduate students who were attracted. Support from cooperating industry and government organizations consisted of both funds and access to field sites and personnel. Cooperative members shared data from their own programs and received data from other members in return.

Using methods borrowed from crop and animal breeding, Zobel developed a thriving breeding program for forest trees. With time, interest in forest genetics grew, and an eastern group, headquartered at North Carolina State University, sought to start its own cooperative. Zobel relocated and helped launch the North Carolina State University-Industry Cooperative Tree Improvement Program. The Texas program continued to prosper and became the highly successful Western Gulf Forest Tree Improvement Program.

As in Texas, the program Zobel established at North Carolina State University combined cooperation and financial support from major forest product companies, state governments, and granting agencies. By 2005, breeding of loblolly pine in the cooperatives, including one established in Florida in 1954, had progressed through the third generation and had produced sufficient seeds for the establishment of 17 million acres of plantations. About 800 million seedlings are being produced each year, enough to establish plantations of 1.3 million acres. Current planting of improved southern pines accounts for more than 50 percent of all forest trees being planted in the United States.

The production of genetically improved plant material is not the only important achievement of the cooperatives. In North Carolina alone, more than 250 graduate students and postdoctoral students have received advanced training in forest genetics and tree improvement. They have been joined by visiting scientists, forest managers, and administrators from national and international organizations in learning about forest genetics through workshops and short courses.

REGIONAL APPROACHES

Soon more universities were training students in forest genetics and tree improvement. Several cooperative tree-improvement programs were developed in various parts of the world. Responsibility for implementing tree improvement was assumed by some governments, including Norway, France, and New Zealand at the national level, and Queensland and South Australia in Australia, Lower Saxony in Germany, and Ontario in Canada, at the state or provincial level. Oxford University took a leadership role in training scientists and fostering forest genetics and tree improvement internationally, particularly for members of the British Commonwealth. Internal programs were established at some corporations, such as Weyerhaeuser Company and International Paper Company in the United States, Aracruz Florestal in Brazil, and South African Pulp and Paper Industries Limited (Sappi) in South Africa.

Economic analyses indicated that the return on investment from tree improvement was greater than from any other forest input. Most of the intensive programs followed the Scandinavian protocol, using seed orchards to produce high-quality seeds relatively inexpensively and in large numbers.

By the 1960s, tree breeders had recognized the importance of provenance and had begun to set up breeding programs to serve particular areas. Seed orchards proliferated, each serving a local region and drawing on genetic variation among local populations and among the individuals within those local populations.

The Scandinavian protocol provided an excellent model for developing the first generation of improved trees. In applying the first step of the protocol, Zobel required cooperators and taught his many students and international colleagues to select only one individual from a stand of trees, to avoid selecting related trees which might lead to inbreeding. In British Columbia, for example, foresters chose the single best tree among each million inspected. To anyone who found two or more neighboring trees that were truly outstanding, it seemed a bit capricious when Zobel insisted that they select only one. But when the offspring of first-generation seed orchards were compared with seedlings from wild seeds collected in some of the original stands, the former usually showed large improvements in lower-trunk and branch form, health, growth rate, and wood properties. Some of the orchard trees' gain in growth rate doubtless reflected avoidance of the inbreeding that occurs in natural stands among related, neighboring trees. Overall, Zobel's approach was vindicated.

PROGENY TESTING

Plus trees are selected from existing plantations or from natural, genetically unimproved tree stands based on their physical appearance, or **phenotype**. The assumption is that a tree with superior physical qualities in the midst of its peers is genetically superior. The unknown is the tree's **genotype**. In simplest terms, any trait of a tree is affected by two components, genetics and environment. The result is explained by the simple formula:

P (phenotype) = G (genotype) + E (environment)

To estimate how much of the phenotype of a tree is explained by genetics, progeny testing is widely practiced.

Progeny testing of plus trees depends on having each parent mated with a similar set of other trees. This can be done with controlled pollination, either using standard mixes of pollens or making crosses with each of several different trees. With wind-pollinated species, open pollination can often be efficient for this purpose, since every parent may be pollinated by much the same pollen cloud. However, wind-pollinated seed has drawbacks for providing future generations of selections.

For controlled pollination, pollen from selected parents is often collected the previous year and stored under special conditions until needed. At the appropriate time, female flowers are isolated within bags to prevent pollination by other trees. When the flowers are receptive, pollen is applied. After the potential for contamination has passed, the bags are removed, and the developing seeds are allowed to mature.

Figure 3H. *Progeny test. A four-year-old progeny test of loblolly pine shows (left) trees representing a family that is free of fusiform rust and (right) trees of a family that is susceptible to the disease. The parent trees in the seed orchard that have the genes for resistance to the disease will be retained for seed production, whereas those susceptible to the disease will be culled. Fusiform rust causes losses of millions of dollars annually by killing young trees and by reducing the wood quality of infected trees that survive to rotation age. Photo courtesy NCSU-Industry Cooperative Tree Improvement Program.*

The mature seeds are germinated, and the resulting seedlings are established in progeny tests of statistically valid design. In these intensively managed tests, the trees are periodically measured for the traits of interest, such as growth rate, form, and pest resistance. The data from the various traits are analyzed to identify the potential genetic gains, and the earliest age at which a trait shows stability is determined. For many important traits, such as growth, form, and wood specific gravity, results from trials of six-year-old southern pines are excellent predictors of the results at harvest; thus, those progeny tests need not be carried to full **rotation**, which may be 25 years or longer, depending on the species.

In review, progeny tests allow tree breeders to rank clones in the seed orchards for genetic merit, and they may contribute some of the better progeny to the genetic base for the next cycle of breeding. The ranking also identifies the

Figure 3I. *Excluding unwanted pollen. The female strobili (conifer flowers) of wind-pollinated species tend to be borne in the upper parts of the trees, while most of the male strobili occur on lower branches. The arrangement increases the proportion of pollen being blown onto the females from other trees, while limiting the amount of selfing. In this 1932 photo, Bill Cumming, a technician at the Eddy Tree Breeding Station, is bagging twigs bearing buds containing female strobili on a ponderosa pine, well before pollen ripens and the female buds open. Photo courtesy of the U.S. Forest Service Institute of Forest Genetics.*

Figure 3J. *A controlled mating. Uncontaminated pollen from a single parent was loaded into this hypodermic syringe. It is being blown onto the opening female strobili, visible through the window near the end of the growing twig. Photo courtesy of the U.S. Forest Service Institute of Forest Genetics.*

Figure 3K. *Mating Earth's tallest parents. In 1979, University of California-Berkeley undergraduate Barbara McCutchan brushes pollen from the world's (then) second-tallest tree (367.4 feet) onto receptive female strobili of the (then) tallest tree (367.8 feet), cloned by rooting mature cuttings so that the work could be done at ground level. The offspring resulting from this controlled pollination have been planted in trials and as specimens worldwide. Interestingly, the members of this redwood family have not grown as fast as many other redwoods in the early decades of these trials, but they may catch up later. Their parents were more than 600 years old when measured. McCutchan went on to earn a Ph.D. in forest genetics at North Carolina State University. Photo by Bill Libby.*

parent clones with lesser genetic worth for removal so that only the better-ranked clones are left to naturally cross-pollinate in seed orchards. Thus, the pollen from genetically superior trees fertilizes the flowers of other genetically superior trees.

PREPARING FOR THE LONG TERM

Seed orchards based on the Scandinavian protocol often gave major genetic gains, and many people considered seed orchards containing only the best progeny-tested parents as representing the culmination of tree improvement. Yet high-yielding modern crops like corn are the result of thousands of generations of selection by farmers, plus more than 100 generations of intensive breeding since around 1900. To achieve cumulative improvement and meet future human needs, forest trees, too, would need more generations of improvement, and because trees are not an annual crop like corn, this would take a long time. Breeding programs have now progressed through the recurring cycle of selection, intermating, and evaluation only two or three generations since the beginnings of the Scandinavian protocol.

Early tree breeders tended to practice superintensive selection of plus trees, choosing only trees that looked near-perfect and then culling them based on progeny test results. This created a big problem. With only 15 or 20 parents in many of these early regional orchards, one could not continue breeding among their offspring for very many generations before matings between relatives would occur. Furthermore, the breeders would be left ill-prepared to cope with changes in breeding goals.

By 1970 tree breeders realized that many more parents had to be selected. Instead of seeking the best-looking tree in a very large area, they started taking the best in, say, 5 acres, with the goal of having several hundred plus trees. To achieve this, some tree breeding programs were effectively started afresh, but without abandoning the original selections and seed orchards.

To achieve improvement and still conserve genetic diversity, the breeder can organize material into a hierarchy of populations. At the top is the production population, which typically comprises the seed orchard parents (perhaps 20 clones per orchard), which can be superseded in new orchards as better parents become available. (In this respect, tree breeding is more like progressive herd improvement in animal breeding than most plant breeding, which produces stable varieties.) Below the production population is the breeding population, which is usually much larger; here, cumulative genetic gain is achieved by building up

Genetic Conservation

In the 1970s, genetic conservation was becoming a fundamental part of general conservation goals and strategies. Setting aside large forest reserves subject to minimal human influences was one strategy to maintain the genetic structures and variation found in native populations. This works well for some tree species, but for others there are problems. As one example, in the absence of stand-replacing fire or other site disturbance, reserves of native loblolly pine are typically invaded and then dominated by various angiosperm species. To compensate for such losses, advances in biotechnology have allowed storage of genetic samples, as tissues, that can be recovered many years in the future to produce clonal copies of the original sampled trees.

Dedicated plantings of random samples of native populations in nonnative environments provide another genetic conservation strategy, particularly if the native population is succumbing to urbanization. Such samples of Monterey pine from its native populations in North America were established in genetic conservation plantings in California, Australia, and New Zealand. However, their open-pollinated offspring are often fathered by pollen blown in from nearby plantations of Monterey pine. Controlling the pollination of such genetic conservation plantings is a complicated and expensive option. Storing wild Monterey pine tissue for later propagation and planting avoids the problem of genetic contamination and is less expensive than trying to control pollination in dedicated plantings, but it means the material is in a static form and not necessarily adapted to a changing environment.

Figure 3L. *A population in need of genetic conservation. The Guadalupe Island, Mexico, population of Monterey pine has been in decline since the 1870s. Goats released earlier by whalers repeatedly exceeded the carrying capacity of the island and prevented pine regeneration from becoming established. A herd of these goats is visible in the center of the picture, among remnant trees of what had been a forest. Besides exhibiting the generally fast growth rate of Monterey pine, this population promises to contribute superiority in stem straightness, wood density and stiffness, resistance to an important disease, and winter cold resistance. The last few native trees are dying. Photo by Bill Libby, 1964.*

frequencies of favorable alleles over generations through successive cycles of selection, intermating, and evaluation of the offspring (usually in field trials).

Underpinning the breeding populations, and possibly numbering thousands of trees or more, are the gene resources, which have the greatest genetic diversity but the least (if any) genetic improvement. They can be used to augment breeding populations. The commercial stands resulting from seed orchards are not usually seen as a source of future breeding material. Rather, gene resources include natural stands, plantations established or managed for this role, seed in long-term storage, and stored clonal material. (See sidebar "Genetic Conservation.") Many variations of this hierarchy of populations are possible. Maintaining such a hierarchy, however, can be very expensive and requires strong, ongoing commitment.

Chapter 4
Clonal Forestry

Propagating a tree vegetatively, traditionally by grafting or rooting cuttings, allows one to produce individuals that are genetically identical to the original. Such individuals are members of a clone.

As we have already seen, clonal propagation serves as a breeding tool when it is used to create seed orchards and facilitates controlled pollination for future generations of breeding. It also can be used to replace or supplement progeny testing, to provide materials for research, and to increase planting stock from top-ranked parents, as is being widely practiced with Monterey pine in New Zealand. Such planting stock from controlled pollination avoids the problems that arise when genetically inferior pollen blows in from outside the seed orchard.

Mass propagation of particular, known clones is termed clonal forestry. For this, grafting is seldom even an option: it is expensive and often beset by delayed graft incompatibility. With many tree species, notably poplars, willows, and Chinese-fir, rooting cuttings is the easiest method of propagation. For them, clonal forestry is the system of choice for plantation forestry; this is one reason poplars have been the subject of genetic improvement programs in various parts of the world. The availability of known, reliable clones, however, can be a disincentive to the quest for still-better clones that can come from a well-structured breeding program. Nursery operators may concentrate on producing the clones that perform best in the nursery rather than those that excel in the field. And a vulnerability, say to a disease or insect, is shared among all members of the clone.

Clones for clonal forestry are usually selected from crosses between the better plus trees of the breeding population, the most promising individual seedlings being chosen to be asexually propagated. The resulting **propagules** are then established in clonal trials, and as with progeny testing, the inferior clones are culled, leaving only the best clones for plantation establishment.

Cloning and breeding are two technologies that feed off each other. To be efficient, for example, a breeding program selects for only a limited number of traits per generation. Among those offspring with above-average expression of the selected traits, a few with outstanding expression of some other valuable traits can be found and cloned. This greatly expands the number of characteristics that can be selected in trees to be deployed in a combined breeding and cloning program.

Figure 4A. *A eucalypt clonal plantation. This Aracruz Florestal flooded gum clonal plantation in coastal Brazil is extremely productive of harvestable wood. These trees are three years on site and already average 60 feet in height. However, this species needs excellent weed control for successful establishment, as well as early growth on fertile soils for such high productivity. Because eucalypts are very sensitive to herbicides, genetic engineering for herbicide resistance in flooded gum clones is an attractive goal. Photo courtesy of Bruce Zobel, 1979.*

To allow large-scale cloning, a breeding program can be adjusted to select for trees more amenable to cloning. At the present state of technology, not all clones can be efficiently rooted as cuttings or propagated through other methods. These properties can likely be improved by breeding, but selecting for them can come at the expense of genetic improvements toward the original breeding goal.

PRODUCTION-SCALE CLONING

Once superior trees have been developed, the next challenge is propagating enough trees to establish large plantations. Until about 1950, almost all forest tree species were either naturally regenerated or artificially regenerated using seedlings. By the 1980s, clonal propagation was being adopted for a substantial and increasing number of species.

Cloning a superior tree replicates it. Unlike seedlings, the cloned offspring are not genetically recombined through sexual reproduction. They therefore do not contain the great variety of new genetic combinations that make seedling offspring all different from each other and from their parents.

Whereas poplar and willow are easily cloned, the cuttings of most other species suited for plantation use need to be grown in a carefully controlled environment until they develop roots. Rooting cuttings was tried for many forest tree species, but attempts to root cuttings from mature trees either failed or the trees thus produced had weak root systems and grew unsatisfactorily. By 1950, it was found that for at least some of the difficult-to-root species, cuttings from young seedlings could be effectively rooted, and they grew well. But the goal was to clone superior trees whose performance was known, not young seedlings whose performance to harvest was unknown.

The reasons why older trees could not easily be cloned became clear in the second half of the 20th century. In several species of trees, it was found that buds that produce branches are not at the same developmental stage in different parts of the tree. Buds formed near the top of the tree are the most mature, while those near the base of the tree, developed decades or even hundreds of years earlier, are the least mature. Therefore, twigs collected as cuttings from near the tree's base or from stump sprouts can generally be effectively rooted. These rooted cuttings develop similarly to seedlings, taking years or decades to attain competence to produce pollen and seeds. Cuttings collected from near the top of the same tree grow to produce characteristics of the upper crown, including immediate sexual competence, but they cannot generally be effectively rooted and are better propagated by grafting.

As maturation state was increasingly understood, techniques were developed to prevent some members of a clone from maturing so that they could serve as future cutting donors; other members of that clone were grown to maturity in evaluation tests. Selections could be made based on performance during all stages of each clone's development to harvest age and beyond, and the selected clones could then be propagated for production plantations from the still-juvenile donors.

There are also some advantages to initiating plantations at a maturation state beyond juvenility—for example, to avoid some juvenile diseases, or to obtain better branch architecture and stem form. Cloning technology that takes advantage of differences in maturation state allows for manipulating trait expression. This option has been employed for centuries in fruit tree orchards, where the goal is to produce fruit quickly without waiting for newly planted trees to grow through the juvenile and adolescent stages of maturation.

In the mid-20th century, new technology enabled researchers to proliferate plant tissue—from shoots, buds, roots, or leaves—on artificial media in sterile conditions in the laboratory. Such **tissue culture** was successful with first a few, then many species of plants. Soon it became possible to induce shoot and root

Figure 4B. *Cloning aspen. Aspen had been difficult to clone by rooting cuttings or by grafting. But both roots and shoots can be cultured to mass-produce whole plants of identical genotype. Organ culture of both shoots and roots has been conducted at Germany's Institute of Forest Genetics at Grosshansdorf. Photo by Bill Libby, 1982.*

development to recover independent plants from these cultures. Multiplication rates increased, and large numbers of a given clone could be reliably produced in a short time. Tissue culture has become the preferred method of clonal production for some species of plants, including several species of forest trees. This technology is also useful for moving clones across quarantine barriers, since the cultures are free of insects and diseases.

Cells in a tree other than the pollen and egg cells are called **somatic cells**. As the 20th century drew to a close, researchers learned to produce and multiply embryos from somatic cells and, ultimately, to **encapsulate** those somatic embryos to produce seeds. Using this technology, huge numbers of plants per clone can be propagated in a short time. Even more important, some of the tissue used to produce the embryos can be frozen at very low temperature and then revived. This procedure allows performance testing of the clone under field conditions while it is held in storage, typically as **embryogenic tissue**. Then, as needed, samples of the embryogenic tissue are withdrawn from storage, greatly multiplied, and the resulting clonal embryos are germinated and used for plantation establishment. Evidence is accumulating that new trees of a superior clone grown from such stored embryogenic tissue reliably duplicate their clone's previous known performance.

Although the problems are far from fully solved, increased understanding and research have allowed the use of cloning to expand. Before 1950, only a few forest tree species could be cloned, but at the beginning of the 21st century, clones could be created from hundreds of tree species.

ADVANTAGES OF CLONAL FORESTRY

Clonal plantations have a number of advantages, both real and potential:

- The relatively uniform traits and performance of the trees allow nursery practices, silvicultural treatments, harvest schedules, processing, and end uses to be tailored to each clone.
- Managers can deploy clones that are known to be adapted to specific kinds of sites, a particularly useful approach when the sites are not extensive enough to justify separate seed orchards.
- Desired characteristics produced by unusual genetic combinations can be replicated more effectively by cloning than through traditional breeding, particularly if the traits involve complex inheritance.

Figure 4C. *Somatic embryos. Somatic embryos of Monterey pine are developing from tissue masses that grew on culture medium after being taken from a sexually produced early embryo. Development of green chlorophyll is just beginning. When harvested at this stage and moved to a different medium, the embryos complete development of green foliage and then grow much like sexually produced seedlings. Unlike sexually produced seedlings, however, they are genetically identical members of a clone. Photo courtesy of the New Zealand Forest Research Institute.*

- Individuals that depart strongly from the usual correlations (for example, between low wood density and fast growth rate) can be replicated (producing trees that both grow fast and produce desired wood properties).
- Clones that perform to expectation on a range of sites, as well as those that are superior for selected site conditions, can be identified.
- Clones whose performance is predictable could be planted in prescribed sequence so that some trees are harvested early and others are grown to final harvest. As the science develops, clones might even be sequenced so that neighboring trees of complementary clones make different moisture and nutrient demands on the site or the same demands at different times.

The advantages of clonal forestry are vividly illustrated by its large-scale adoption during recent years in some species that have been traditionally propagated as seedlings. Large areas of extremely productive clonal eucalypt plantations are now growing in Brazil, South Africa, and some other tropical and subtropical countries—more than 1 million acres in Brazil alone. Extensive clonal acacia plantations have also been established recently in Indonesia and China.

Clonal plantations may eventually achieve something akin to the success of certain cereal crops whose yields have been boosted by selecting for reallocation of biomass. Cereal crops have been bred to have shorter stems so that less biomass ends up as low-value straw. This genetic improvement substantially increases grain yields even though the biomass per acre remains the same. In a cereal, the uniformity of the crop makes it possible to take advantage of characteristics that boost harvest efficiency. In trees, cloning can give that uniformity, with the promise of some similar benefits.

At least some tree species need only half or less of their normal complement of roots for support and for taking up water and nutrients. The remaining roots engage in a sort of quiet underground warfare to occupy territory and thus deprive neighbors of water and nutrients. Researchers may find trees that allocate less biomass to roots and more to stem growth.

Above ground, forest trees compete for light by growing tall and keeping their crowns at least as high as those of their neighbors. But if all trees in a plantation are known to have limited height growth, resources that went into upper-trunk wood might be reallocated to the more useful and thus more valuable lower logs.

Many tree species allocate substantial biomass to reproductive organs. But domesticated clones don't need to reproduce sexually. If clones can be engineered to delay or eliminate sexual reproductive competence, some or all of

Figure 4D. *Addressing a clonal plantation disaster. Poplar clone I-214 is a Euramerican hybrid that combined the good growth of eastern cottonwood, native to the United States, with the good rooting of black poplar, native to Europe. Produced and tested in the 1940s in Italy, it was soon planted in enormous numbers in many temperate climates around the world. In Serbia and Croatia, for example, a pulp mill was built with expected production based on these I-214 plantations. But by 1980, the clone was seriously affected by two known poplar diseases and a previously unknown virus. As it became clear that large monoclonal plantings of I-214 was a bad idea, the Forest Research Station at Izmit, Turkey, pursued a novel and aggressive strategy. Korhan Tunçtaner, shown here in 1992 with some of his collection, gathered thousands of seedlings from much of cottonwood's native range. Ten cuttings were set from each seedling. Based solely on effective rooting, the best 10 percent of these clones were then deployed to field tests. By 2003, 30 eastern cottonwood clones had been identified that rooted satisfactorily and grew as well as or better than I-214. The best of them, 74-047, produced 55 percent more wood per acre than I-214. Production could then be maintained or increased with multiclone plantings. Photo by Bill Libby.*

this biomass could also be reallocated to lower logs. The slogan for this approach is "Make wood, not love."

RISKS OF CLONAL PLANTATIONS

By the 1980s, using clones in plantations was sufficiently promising that several kinds of risk were being given serious attention.

Loss of genetic diversity. If very few clones are deployed over wide areas, the stands are vulnerable to catastrophic loss from a single physical or biotic event, and particularly so if some insect pest or pathogen adapts to one or more of those clones.

Simplified ecosystems. Agriculture is efficient and highly productive not only because of genetic improvements in domesticated varieties, but also because those varieties are grown in simplified ecosystems. Most people accept that, but forest ecosystems are another matter. Many professional biologists, as well as some in the general public, oppose converting complex forests into so-called monocultures by intensively managing areas planted with a single species, even though large "pure" stands often occur naturally. Large areas planted to a single clone raise even more concerns.

Increased mutations. In tissue culture, certain practices have produced new sorts of genetic variation at rates far higher than for typical mutations. This **somaclonal variation** within clones has been disquieting both within and outside the forestry community.

Unintended consequences. During the 1970s, there was increasing concern worldwide about the manipulation of nature, and some of the unintended consequences of some well-meaning manipulations were widely publicized. Substituting clones for seedlings seemed a particularly high-handed attempt to control nature, and opposition developed.

The structure of the genetic diversity that protects a plantation of trees against epidemic outbreaks of insects or diseases is now better understood. The crucial point: the number of genetically different clones or trees in a plantation is not as important as how different the clones or trees are from each other. Deployment of fewer than five clones over large areas is now considered a management error rather than an unavoidable risk. However, in a plantation of 14,000 trees, a mixture of about ten unrelated clones, each in 1,400 copies, would usually be less at risk of an epidemic than if those 14,000 trees were of seedling origin, with each seedling genetically unique but in large part closely related to each other.

Even though planting ten unrelated clones reduces biological risk, the question remains whether those clones should be intermixed within an area or planted in single blocks. One needs to anticipate that one or more of the clones will fall victim to a pest or climatic event. Salvage harvesting of trees of one clone in an intermixed plantation may not be feasible without damaging adjacent trees of healthy clones. Moreover, the land occupied by the affected clones may remain unused until the remainder of the plantation is harvested. Single-clone block plantings, on the other hand, allow salvage cutting and permit the land to be used as soon as the affected clone is removed. Operational experience with

From Biological Deserts to Complex Ecosystems

The ecosystems that develop in intensively managed clonal plantations cannot duplicate those in virgin old-growth forests. It is appropriate that adequate areas of old-growth forest be reserved or restored in most or all forest regions. If that is done, then young-forest ecosystems don't replace biodiversity on a landscape level; they add to it.

In New Zealand, prior to the development of tree improvement, it was necessary to plant about 1,000 Monterey pine seedlings per acre in order to have about 100 good trees to harvest. Those closely spaced young trees so fully occupied the site with their roots and crowns that critics called such plantations biological deserts. Most of these Monterey pines had unacceptable form, growth rate, or health, and only about one in ten developed the combinations of desired characteristics that made for a reasonably valuable tree at harvest. During plantation development the other nine in ten died from health problems or competition, or were thinned out as their problems became evident and to give growing space to the desired "crop" trees.

After reliable, highly selected clones became available, only 200 to 300 trees needed to be planted per acre to ensure 100 desirable trees at harvest. This not only saved time and money, it left open niches for other organisms on the site. Many shrub, fern and herb species have established themselves among the widely spaced young trees, and many maintain themselves as the plantation ecosystem develops. Insects, birds, and other species that live on or in those plants become part of that complex ecosystem. In extreme cases, such diversity could constitute a major weed problem, but here, the use of reliable clones can improve biological diversity in plantations by using wide-spacing designs.

Figure 4E. *Vigorous understory. Highly variable seedling performance has required managers to compensate by planting many more seedlings than would be needed to produce the final crop. Using reliable clones, fewer trees need to be planted at stand establishment. With fewer trees, more light enters the stand, and a more diverse understory appears, one that can develop during the life of the forest. The interior of this near-mature radiata pine stand shows vigorous undergrowth, mainly ferns and grasses. Photo courtesy of the New Zealand Forest Research Institute.*

eucalypts and acacias, and to a large extent with poplars, favors a mosaic of single-clone block plantings.

Regardless of the clonal plantation design, if only one or two forest tree clones susceptible to some epidemic were widely deployed for many years, plantations of those clones could suddenly and massively fail, and they could not be quickly replaced. Such a threat faced American corn growers in 1970, when for about two decades, a high percentage of cornfields had grown varieties of corn that all shared a genetic feature useful in controlling pollination; however, that feature made them susceptible to a strain of corn blight. Then, weather favorable to the blight supported an epidemic that swept through most of the corn-growing region. But the genetics of corn had become well understood, and corn-breeding programs had reserves of resistant varieties. The lesson for forestry is to better understand the genetics of important plantation species and to maintain a ready reserve of diverse genetic material.

Recognizing the need for risk spread among clones in forestry, various European countries have imposed regulations (which are often ultra-cautious) that prescribe minimum numbers of clones to be used as well as intermixing requirements. Clonal forestry clearly has its risks, but they can be much outweighed by the advantages if the risks are recognized and duly managed.

Chapter 5

The Science of Genetic Engineering: A Primer

The **domestication** of tree populations began, as we have seen, with relatively traditional plant biotechnologies. Some techniques have long been practiced in horticulture and arboriculture, including grafting, deliberate cross-pollination, and raising planting stock in nurseries. Mid-20th-century advances in biotechnology included the use of plant-growth regulators to improve rooting of cuttings.

For many people, *biotechnology* implies the use of modern **genetic engineering**, but it is much more. The term, biotechnology, came into common usage in the 1980s, and its several definitions continue to change. Broadly defined, it is anything that combines biology and technology. One may refer to "modern" biotechnology, which postdates the discovery of the structure of deoxyribonucleic acid, or **DNA**. Modern biotechnology, while including the analysis and manipulation of DNA and the artificial **insertion** of DNA fragments into organisms, also includes other recent and demanding technologies, such as tissue culture and embryogenesis to produce plants in laboratories. Much of the new biotechnology is dependent on recent refinements and revolutionary developments in the science and techniques of biochemistry, molecular biology, and computer science. Along with the advances have come some formidable challenges—not only technical challenges, such as how to apply these advances, but also ethical concerns and environmental issues.

Biotechnology has become more controversial as the level of the technology has increased. Cloning of fruit trees by grafting is "low-tech" biotechnology, long established and not very controversial. Cloning of frogs, cows, Dolly the sheep, pet dogs, and potentially, a human—by producing cloned embryos in the laboratory—is "high-tech" and even shocking. Similarly, even though startling numbers of genes are possessed in common between very different organisms, transferring single genes from one species to another is, for some people and organizations, anathema. Even genetically modifying forest trees so that they are better adapted to difficult climates, or so that they sequester more carbon, is causing concern regardless of the potential benefits.

THE DNA REVOLUTION

The year 1953 marked a breakthrough that was catalytic in the development of modern biotechnology: the discovery and subsequent understanding of the structure of DNA by James Watson, Francis Crick, and their several colleagues and competitors.

The implications of DNA's structure as revealed in 1953 were vast. It could account for the puzzling ability of genetic material to replicate itself accurately, and it gave the clue as to how such an apparently simple molecule could generate the gigantic complexity of living organisms. The linear structure also fit well with the existing knowledge that genes are contained in chromosomes.

The precise nature of differences between alleles also became clear with the discovery of DNA: it was determined to be the result of differences in one or more of the bases. An allele is important if it affects an important function in the organism, typically by either coding for a different **amino acid** and thus a different protein or altering how much of the protein is produced. Until this information was obtained, the closest geneticists had generally come to identifying the source of allelic differences was recognizing variants of several enzyme proteins. These variants, called either isozymes or allozymes, remained an important laboratory tool in many genetic studies for about 25 years, from the 1960s until DNA technology was well developed.

Once the genetic code was deciphered and DNA's base sequences could be read efficiently, it became clear the code worked the same in most organisms. That was a crucial finding for the development of genetic engineering. With improved laboratory equipment, it became possible to read DNA's base sequences progressively faster. The number of functioning genes in a typical plant is about 30,000, and storing, organizing, analyzing, and interpreting the massive amount of information being obtained was integrated in a new scientific discipline called bioinformatics. Further progress in understanding how genes function allowed genes not only to be identified, but also to be extracted or assembled artificially and massively copied.

Some of these advances are of special significance for long-lived organisms like forest trees. For example, it is possible to identify genes and some of their functions without having to follow inheritance patterns over two or more decades-long generations, as needed to be done in 20th-century genetics studies.

APPLICATIONS

A spectacular success for forest biotechnology already exists in the application of DNA "fingerprinting," much as is now used in criminal investigations. It allows the genetic identity of individuals to be verified. In plantation forestry, this application is proving embarrassingly successful because past misidentifications are now being found and corrected. As tree breeding programs advance,

The Structure of DNA

DNA usually exists as pairs of strands that are wound like a twisted ladder, making a double helix (5A). The "backbone" of each strand, in which sugar and phosphate groups alternate, is on the outside. Connecting the strands in the middle, like rungs of a ladder, are four groups of atoms called **bases**: thymine (T), adenine (A), guanine (G), and cytosine (C). These bases form a four-letter "alphabet" to provide genetic information. The T of one strand is joined to an A in the other strand, as are G and C. The two pairings each take up the same space, allowing the configuration of the double helix to be uniform despite differences in the shapes of the four bases.

One DNA strand, from which information is transcribed, is called the sense strand, the other the antisense strand. The sense strand can also be a template for forming a new antisense strand, and vice versa, thus allowing a DNA molecule to copy itself (5B). Each strand can serve as a template for forming new and identical double-strand molecules. Under the right laboratory conditions, known "tagged" single strands will find and pair with complementary strands from a donor organism. This allows researchers to determine whether a known sequence of DNA is present in a prepared specimen.

The DNA of an organism consists of the coded genes plus "noncoding" DNA. The amount of noncoding DNA is variable, sometimes constituting a high proportion of the total DNA. Its role is still uncertain, but being variable, it can be used for genetic fingerprinting.

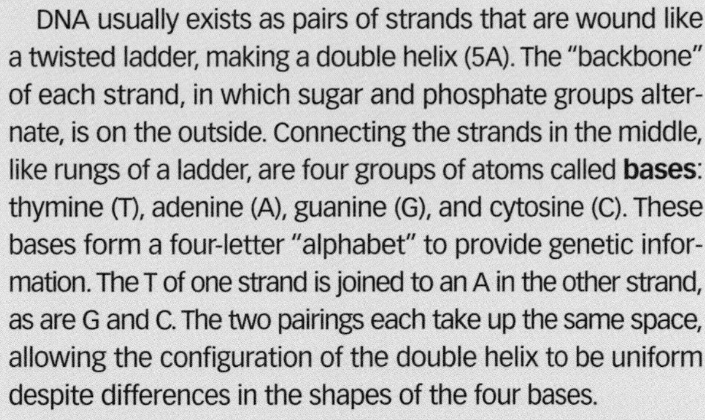

Figure 5A. *Double helix.*

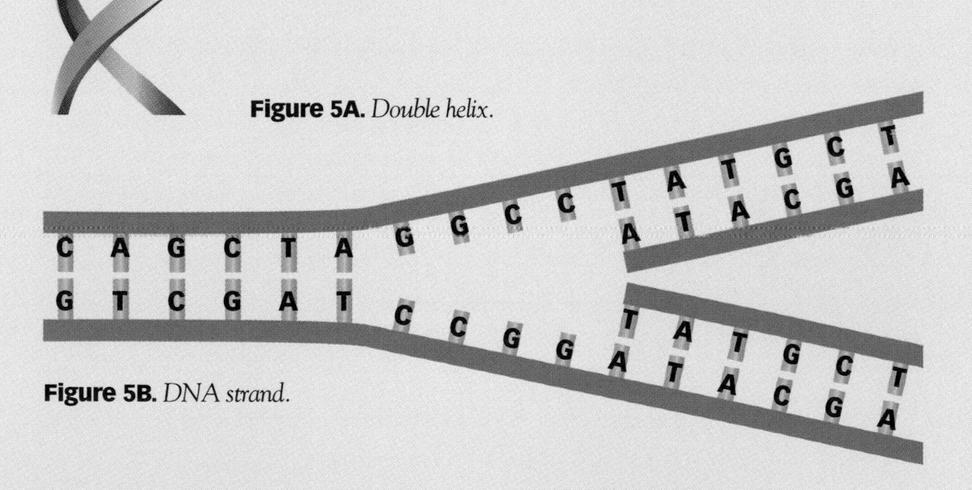

Figure 5B. *DNA strand.*

accumulated purposeful genetic change could be lost or misapplied through misidentification of families, cell lines, or clones. Indeed, DNA fingerprinting becomes progressively more important: it not only allows the identity of clones to be verified, it also makes it possible to detect pathogen infection.

DNA analyses are also proving valuable in characterizing the levels and organization of genetic variation within and among populations of trees. In a less practical but still very important application, DNA sequences are powerful tools for clarifying the evolutionary relationships among species and among different populations within species. These important biotechnology tools enhance our understanding of plant genetics without involving genetic modification, and they can lead to knowledgeable decisions about threatened or endangered species.

Analyzing DNA variants and their biochemical consequences makes it possible to track DNA action in two directions. In one direction, observable variations in traits can be traced to the regions or particular loci within chromosomes that genetically specify them, and even to specific short sequences of DNA within a gene. In the other direction, coupled with other biochemical techniques, the pathway leading from a gene's information to actual tree traits can be traced.

GENETIC TRANSFER AND TRANSFORMATION

Once it is known what certain genes can do and where they are available, it is possible to purposefully transfer genes between organisms. Alternatively, the activity of genes already present can be usefully altered. Any inserted DNA sequences, whether complete genes or only parts of genes, are referred to as **transgenes**. The transgenes can be of natural origin, or they can be artificially synthesized. Also, more than one gene or part of a gene can be transferred at a time; this is called stacking.

Several ways have been developed to insert genetic material into the target organism. One such **transformation**, as the process is called, relies on a natural process: the crown-gall bacterium *Agrobacterium tumefaciens* causes infected plants to produce galls by inserting some of its DNA into the host plant's cells. Genetic engineers have taken advantage of this natural genetic engineering system. After deliberately inactivating the unwanted genes that the bacterium inserts to form galls, they use this genetically modified bacterium as a way to insert desired transgenes into plants.

A "gene gun" can also be used for "biolistic" insertion of transgenes leading to genetic transformation. The target tissue is bombarded with tiny DNA-coated gold spheres propelled by compressed gas. Some of these spheres by

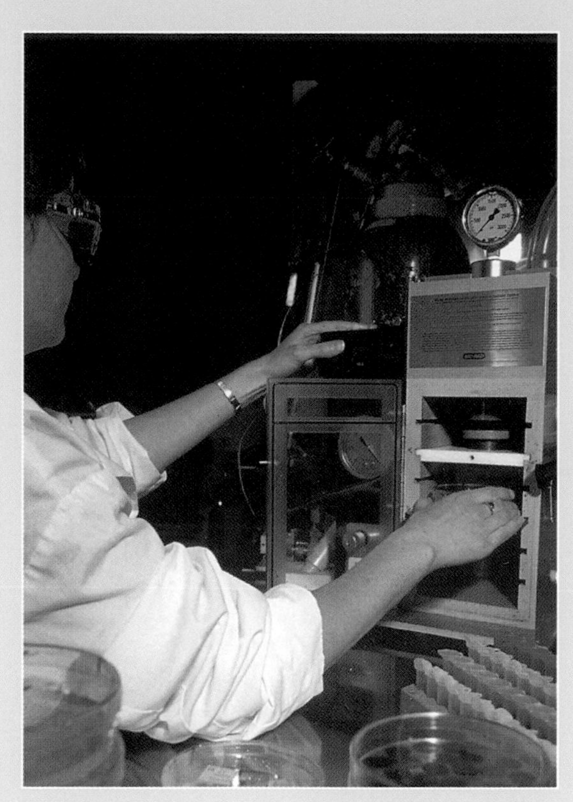

Figure 5C. *Inserting DNA. In "biolistic" genetic transformation, a gene gun is used to bombard the target tissue with gold spheres coated with the desired DNA, using a highly compressed gas. If the DNA is, by chance, incorporated in a chromosome, the transformed cells are grown on into complete plants; then screened for stability and expression of the introduced transgene(s). Another, perhaps more natural way to insert a transgene into a chromosome is to use the bacterium* Agrobacterium tumefaciens; *which works well in some species. Photo courtesy of the New Zealand Forest Research Institute.*

chance arrive where their DNA can be incorporated in a chromosome. The transformed cells are encouraged to grow and differentiate into complete plants.

What scientists are doing is actually not unknown in nature. In the course of evolution, nature has performed many feats of genetic engineering, some of which vastly transcend what scientists are now attempting. In many plants, genes relocate spontaneously to different loci within the same chromosome or even to different chromosomes. As well, viruses and some bacteria can insert and integrate specific segments of their DNA into their hosts' DNA, and thus get those host cells to work for them.

Because DNA is usually inserted at essentially random sites within the chromosomes, the process has the potential to disable or change some of the resident genes. The expression of the transgene may prove to be transient or unstable, and there can be unexpected interactions between a transgene and resident genes or other transgenes. Most of these uncertainties can be addressed during laboratory screening and field testing; those newly transformed clones with stable integration and satisfactory expression of the introduced transgene are further developed and propagated.

In annual agricultural crops reproduced by seeds, one must usually ensure that the desired transgene is present and functioning in chromosomes from both male and female parents. Only then will such plants "breed true" and the transgene effect appear in all the offspring. In contrast to various agricultural crops, many forest tree species can now be effectively cloned. Once a new trait has been introduced to a single tree through genetic engineering, that tree can be propagated by cloning, leaving no need for it to breed true. Thus the transgene needs to be successfully incorporated in just a single plant, which can then be cloned.

Chapter 6

Tree Improvement at the Start of the 21st Century

The discovery of DNA structure in 1953 gave impetus to efforts to use genetic knowledge in ways that would improve crop yields and human health, and today genetic engineering is a promising complement to conventional breeding in agriculture. The techniques of genetic engineering have already been applied with cultured bacteria and yeasts for production of pharmaceuticals and various other products, and deployed on large scales with corn, soybeans, rice, and other agricultural crops.

As we have seen, conventional breeding is a form of genetic engineering: it is based on cycles of selection and **recombination** of genetic variation. Continued over successive generations, this leads to increases in the frequencies of favorable alleles at the various loci that affect desired characteristics. It thereby incrementally achieves desired genetic modifications. However, this approach has some important limitations.

Conventional breeding mostly concentrates favorable alleles that are already present in the local population or breeding **line**. New alleles may arise by mutation, but favorable new mutants are rare. Desired alleles may also be acquired by hybridization with other provenances or species, but this approach can carry much unwanted genetic "baggage."

When parent trees are mated by the breeder, half of the alleles of each parent contribute to each offspring, and with rare exceptions, no two seedlings get the same set of alleles from either parent. Each of these unique gene packages, contained in the pollen or the egg, will doubtless include many alleles that are not wanted by the breeder. Some of them may be at loci physically close to desired alleles on the same chromosome, making it particularly hard to select out those unwanted alleles. Although selective breeding will eventually eliminate most of the unwanted alleles, this process requires many generations.

Genetic engineering, on the other hand, involves the insertion of one or more short sequences of DNA into recipient cells. Researchers can thus insert genes not naturally found in the species, or insert a different allele of an already-present gene without bringing along all the genetic baggage. This is particularly useful if the desired allele comes from a different provenance or species that is poorly adapted to the site where the genetically modified trees will be planted.

TRANSGENIC TRAITS FOR FOREST TREES

The potential benefits of genetic engineering for forestry are conceivably huge, and researchers are concentrating on conferring properties that are slow or impossible to obtain by conventional breeding. Traits controlled by one or very few "major" genes tend to be targeted.

Herbicide resistance has become a common goal when modifying crop plants, from corn and soybeans to plantation trees. The technology for conferring it is now well developed. It is particularly attractive for poplars and eucalypts, which can be difficult to establish if they face competition from weeds. Most **hardwood** trees are naturally sensitive to almost all herbicides. Furthermore, many require high soil fertility, which also favors weeds. Trees that are modified to be resistant to herbicides, however, can grow very quickly if herbicides can be used for weed control.

Herbicide resistance can also have major environmental benefits. It can eliminate the need for soil cultivation or fire, both of which can increase soil erosion and cause nutrient losses. For good regeneration of forest crops, such resistance facilitates the use of herbicides like glyphosate (Roundup®), which leaves no persistent residues.

The unintended development of resistance to herbicides by exposed weeds is less of a concern in forestry than for herbicide use with annual agricultural crops. In timber plantations, herbicides are applied only during the first year or so before and after planting—a small fraction of the usual rotation. That makes natural selection for resistance in weeds relatively ineffective, since the weeds are exposed to the herbicide only a few years out of many, rather than every year.

Insect pest resistance has been conferred, for example, by inserting a gene from *Bacillus thuringiensis*, a naturally occurring soil bacterium, to produce a toxin that is highly effective against defoliating larvae of moths and butterflies. Incorporation of this *Bt* gene is now commonly used for some agricultural crops, notably corn and cotton, and is being evaluated in a few forest trees, mostly poplar hybrids. An operational planting of poplars transformed for resistance to these defoliating insects was reported from China in 2003, when about 2,000 acres were planted; additional acres have been planted each year since then.

Disease resistance may also be achievable through biotechnology. (See sidebar "Chestnut Blight and Genetic Engineering.")

Figure 6A. *Monterey pines susceptible and resistant to herbicide. Because certain species are easily outcompeted by weeds, a herbicide applied soon after planting can help the young trees get established. The plants on the left are without transformed herbicide resistance; those on the right were genetically engineered for resistance to Buster®* (glufosinate ammonium). *Plants are shown following the application of the herbicide. Photo courtesy of the New Zealand Forest Research Institute.*

Wood properties, such as grain orientation, strength, and lignin content, are other targets for genetic manipulation. Because these properties vary from tree to tree and are highly heritable, genetic gains have already been obtained from conventional breeding programs, but genetic engineering holds the promise of further improvement.

Genetically engineering wood properties of some species into other species could deliver useful products. For example, in hardwoods, the lignin that binds cellulose to make wood rigid is more easily dissolved for pulping than the lignin in conifers. Whether genes that program easily dissolved lignin can be inserted in conifers without unacceptable biological or technical side effects remains to be seen, but if so, efficiencies of pulping and effluent management would be greatly increased.

A more distant goal would be to change some wood properties to better suit completely new uses, such as growing wood—a renewable resource—as a feedstock for material currently produced from petrochemicals, or to make liquid fuels.

Environmental tolerances are yet another way to improve agricultural and forest plants. If trees were genetically modified to tolerate salinity, for example, they could be grown in areas where soil salinity has become a problem for agriculture.

Chestnut Blight and Genetic Engineering

Most genetically engineered trees will likely be deployed in plantations, but an important exception may occur in a project now under way in the United States.

In the early 1900s, a blight pathogen introduced from East Asia swept through eastern North American forests, mildly infecting many oaks and destroying almost all the American chestnut trees—a stately species that provided food for many species of wildlife, very hard and durable lumber for construction, and handsome furniture-grade wood for cabinetmaking. In the following decades, healthy, mature American chestnuts were found in remote areas, but they always succumbed when exposed to the pathogen. The disease kills only above-ground portions of the tree, so chestnut sprouts continue to appear to this day; they grow for a few years, and then become infected and die. The oaks provide a reservoir for the disease even in the absence of mature chestnuts.

Many approaches to overcome the disease have been tried. American chestnut seeds were even exposed to radiation in the hope that mutations conferring resistance would occur. None were found.

The disease in China coevolved with Chinese chestnut, the native chestnut species in that country; thus Chinese chestnuts have substantial resistance to the blight. A conventional genetic breeding program of hybridizing American chestnut with Chinese chestnut has produced hybrids with some resistance, but compared with American chestnut they are poorer in stature, wood quality, and quality of their nuts. The American Chestnut Foundation began a systematic backcrossing program between the two species to achieve a tree that is 15/16 American chestnut, with the Chinese chestnut genes supplying good but still variable resistance. Additional generations of backcrossing may be needed to further dilute the undesirable characteristics of Chinese chestnut while maintaining its disease-resistance genes.

High-tech genetic engineering provides two opportunities for restoring American chestnut to its original ecological niche: (1) to identify the genes in Chinese chestnut that confer resistance to the fungus and use that information in a directed breeding program to consistently produce offspring with the desired gene combination for resistance, and (2) to insert into pure American chestnut the few Chinese chestnut genes that confer resistance without incorporating the genes that adversely affect tree form, wood quality, and nut production. Building on the series of pedigreed backcrosses, researchers have located the regions of DNA responsible for resistance to the blight. Work is now in progress to identify the precise genes.

If this genetic engineering feat works, an important question may be whether these genetically engineered American chestnut trees should be introduced into eastern U.S. forest ecosystems. If they are, and if they thrive, not only will the chestnut component be restored to those ecosystems, but people in the 2050s and later can once again, as did their ancestors, enjoy chestnuts roasting on an open fire.

Figure 6B. *Return of the chestnut? Paul Sisco of The American Chestnut Foundation stands beside a backcross tree of American chestnut and Chinese chestnut that was planted at the Biltmore Forest School in the Cradle of Forestry Discovery Center (Pisgah National Forest, North Carolina), on the occasion of a 2005 colloquium attended by U.S. and European foresters. Photo by Steven Anderson.*

Or trees could be developed with the ability to sequester and remove heavy metals from contaminated soils. Greater tolerance of drought or growing-season frosts are other useful and apparently attainable goals. Genetic engineering, then, may provide a viable tool to address the implications of potential global climate change.

Reproductive sterility entails suppression of pollen and seed production—something that is essentially impossible to obtain through conventional tree breeding. Sterility in plantation trees offers two benefits. First, it may divert a tree's nutrient resources away from sexual reproduction and into greater wood production. Second, it prevents transgenes from spreading into adjacent ecosystems. Even though sterility of transgenic trees would appear to be highly beneficial, there is uncertainty whether it could ever be fully achieved, and some debate over whether it would have higher priority than other transgenic goals.

TECHNICAL AND MANAGEMENT CHALLENGES

Development of any new technology poses challenges. With genetic engineering, appropriate DNA insertion must first be achieved. Then the inserted transgene must remain integrated in the host DNA. The inserted transgene must function as intended, with no unacceptable damage to resident genes or adverse side reactions. Promising transformed clones that do well in the laboratory must then pass extensive field tests for performance and demonstrate freedom from unacceptable side effects. Finally, perhaps many years later, a complete technology chain must be developed to allow mass-scale production of successful genetically engineered clones.

Managing the development and practical applications of genetic engineering involves many choices: the technological pathways to be pursued, the traits to be improved or introduced, the balances to be struck between new and previous technologies, and the personnel to direct the applications and their integration. A coordinated program requires understanding of both conventional breeding and high-tech tree improvement and the appropriate roles of each. Altogether, genetic engineering poses major organizational and technical challenges.

POTENTIAL RISKS AND CONCERNS

Many adverse human health and ecological consequences have been cited as reasons why the technology of genetic engineering should not proceed. The potential hazards are many. Some of these issues are new; others arose with plantations and clonal deployment but are now amplified.

Figure 6C. *Clonal testing. Sites containing Norway spruce clones, with two-tree plots, were part of a test conducted by the Lower Saxony Forest Research Station, Escherode, Germany. Such tests are essential before widespread deployment, and they cannot be hastened by currently available technology. Photo by Bill Libby.*

Adverse side effects on crop performance. Sometimes additional DNA is inserted along with the desired DNA. Resident genes could be damaged or inactivated by the insertion process. The inserted transgenes could have unforeseen side effects which could be difficult and slow to identify.

Subtle, delayed-action effects therefore remain a possibility. This hazard may not be too serious for annual crops because it is often possible to switch rapidly to different varieties. For forest trees, however, an effect such as a delayed loss of health could be catastrophic, particularly if it involved a high proportion of plantings made over many years in a region.

For the time being, only a few transgenes are likely to be inserted together. However, if stacking multiple genes becomes easier and many different transgenes are inserted into the same clone—say, one each from a bacterium, a fungus, a fish, an insect, a fern, a duck, and a palm tree—the number of novel combinations in which such transgenes could interact greatly increases.

Effects on human food. Forest trees grown in plantations as timber crops are generally not sources of human food, so food safety is far less an issue than for agronomic food crops. Food containers made from wood fibers of genetically engineered trees could conceivably have an adverse effect on human health, but that seems very unlikely because of the chemicals and heat treatments used in the pulping and papermaking processes. The use of leaves, fruits, seeds, or wood fiber as components of animal feed presents a situation parallel to the feeding of genetically engineered corn to animals destined for human consumption. Some direct food hazards could exist, but verified cases are lacking. Among the various food safety issues, those from genetically engineered forest trees appear minor.

Resistant pests. When genetic engineering is used to confer disease or insect resistance, it is possible that the pest will develop resistance to the inserted gene; a more virulent pathogen or robust insect could be the result. Genes with large effect are generally used, and such "major" genes are often the very ones that can most readily be defeated by subsequent evolution of the pest. Interestingly, however, the insect resistance engineered using the *Bt* gene is so far proving surprisingly durable.

Enhanced invasiveness. Invasive species displace native plants and associated fauna, and some can affect water yields. Some nonnative tree species can be strongly invasive—strawberry guava in Madagascar is an example—but many of the worst offenders are horticultural escapes. Forest plantation escapes that cause problems are at least candidates for harvesting, just as they would be in plantations. Since invasive trees tend to spread via seeds, conferring sexual sterility through genetic engineering would greatly reduce or remove the natural invasiveness of these exotic tree species.

Harmful effects on biodiversity. Rather than presenting a new kind of risk, genetically engineered trees will probably have only an incrementally different impact on biodiversity in plantations. For instance, control measures already used, such as herbicides or pesticides, may unintentionally harm pollinators, beneficial microflora and microfauna, and other nontarget organisms. Control of weeds and pests is already common practice in intensively managed plantations. Where the availability of genetically engineered trees increases the area and intensity of plantation management, the biodiversity on those sites might be reduced or at least altered. However, plantations with greater harvest productivity per acre make it possible to increase the area that can be set aside in natural reserves, thus creating a potential net gain in biodiversity.

Gene escape. Pollen can travel modest to long distances from plantations into surrounding native forests. If the pollen from transgenic trees fertilizes native trees, hybrids can result. This is termed genetic augmentation, genetic contam-

Figure 6D. *A mushroom-like fungus. Fruiting bodies of the fungi genus,* Suillus, *are shown growing in a beneficial relationship with Monterey pine. Such relationships are common: the tree supplying the fungus with sugars and other photosynthetic products, while the fungus provides the tree with mineral nutrients (especially phosphorus) and possibly water. Such fungi are, in many parts of the world, a commonly collected and prized food. If the mushrooms are collected in areas associated with genetically engineered trees, there are conceivable but as-yet-unproven health hazards. A more likely and known risk of eating boletes is confusing poisonous species with edible ones. Photo courtesy of the New Zealand Forest Research Institute.*

ination, or genetic swamping of the native population, depending on the situation, one's point of view, and the extent and consequences of the immigration.

Genetic contamination of natural populations through interpollination is rightly a major concern, especially near genetic conservation reserves. It has long been a problem when incorrect provenances or conventionally improved breeds are used in plantations near native populations and could be a serious new risk with transgenic trees. Both the owners of the transgenic trees and the owners of natural forests and plantations certified to be free of genetically modified trees could suffer legal liability and financial risk if a gene escaped into their stands. Sexual sterility engineered into the planted trees would eliminate this risk.

Horizontal transfer of transgenes. Dire ecological side effects have been postulated as a consequence of the "horizontal" transfer of transgenes—that is, the escape of genes from genetically modified trees to other species. Hypothetically, horizontal gene transfer could occur in various ways. For example, the *Bt* gene that controls defoliating butterfly larvae could conceivably escape from plantation trees to milkweed, the main host plant for the monarch butterfly, leading to declines in this benign insect. That the possibility of such ecological mayhem is speculative also makes it hard to disprove. To cause real ecological damage, horizontal transfer would require a chain of conditions: it must first occur, it must then confer a competitive advantage on the recipient organism, and finally, that advantage must lead to real ecological mischief—for example, **extirpation** of a species or degrading of important ecological functions.

RISK MANAGEMENT

Risk management is an issue for all plantation forestry, which already has to address elevated risks associated with deployment of clones and now must address those of the various applications of genetic engineering. Risk can be measured roughly as the magnitude of a potential loss multiplied by the probability that that loss will occur. How a risk should be managed is influenced by the perceived levels of these two components. In plantation forestry, risks range from high probabilities of minor losses, such as mild localized disease in a few plantations, to low probabilities of catastrophic losses, such as massive failure of a region's forests.

The risks of genetic engineering can be grouped into two main categories: contained risks, which primarily involve the tree-growing operation; and uncontained risks, which involve environments and ecosystems beyond the planting of genetically engineered trees.

Contained risks can be addressed by choosing insertion procedures for transgenes that are less likely than alternatives to cause subtle genetic damage to resident genes. Then, after screening in containment facilities, field testing of genetically modified trees is essential. For example, it is tempting to produce trees with reduced- or modified-lignin wood that is more efficient to pulp, but testing of such radically different transgenic trees should probably last longer than conventional progeny testing, at least in the development stage. Such tests would primarily address tree strength and performance but also monitor ecological ramifications.

The risks associated with unforeseen side effects are important because delays in identifying problems could make those problems much more serious. Risks involving unlikely but potentially dire events, comparable to the corn blight epidemic (see "Risks of clonal plantations," in Chapter 4) could be addressed by risk spread. In clonal forestry, although any one clone may be disastrously susceptible to, say, a new strain of a fungal disease, it is unlikely that all clones would be similarly vulnerable. Thus, multiple unrelated clones are deployed. The challenge is to extend the same principle of risk spread to genetic engineering by using different transgenes for a particular purpose and using separate insertions of those transgenes.

Uncontained risks include genetic contamination of other tree populations, possible effects on food chains, and effects on species composition and ecosystem functions. One strategy may be to locate plantings of sexually competent transgenic trees in areas where escapes of pollen or seed should have minimal effects. However, since tree seeds are dispersed by birds, wind, and water, and since wind-blown pollen of species like Scots pine and Douglas-fir can travel hundreds of miles, the buffer areas around plantations of genetically engineered trees would have to be enormous.

Mitigating factors are that pollen from one location is likely to be out of synchronization with the flowers of the same or related species in another place, and that pollen transferred long distances is likely to be nonviable because of desiccation and solar radiation. Research is badly needed to determine the polluting effect of long-distance pollen spread, but until that information is available, a safe approach to the problem of long-distance pollen dissemination is to deploy trees that have been genetically engineered to be sexually sterile.

A frequent suggestion for managing uncontained risks is the precautionary principle. Severely interpreted, this principle entails refraining from using a technology if there is any doubt about its safety. Such an alternative has its own risks: if vastly greater quantities of wood are needed than can be produced by conventional means, doing nothing now risks catastrophe in a few decades for many people, as well as for the natural forest ecosystems that they, in desperation, will raid.

Although purposeful genetic modification can serve to counter some known risks, such as existing diseases, it can also incur various risks that are essentially unknown. The greater the genetic gain sought, the greater the unknown or uncertain risks tend to be—much as in financial investments.

SOCIAL ACCEPTANCE

Despite continuing controversy, genetically engineered crops in agriculture, including corn, wheat, and soybeans, are now a well-established reality, and their use continues to grow. As of 2004, about 200 million acres had been planted to transgenic crops worldwide in 18 countries. Genetic engineering as it applies to trees is a more recent activity, and though still largely in the research and development phase, transgenic trees are likely to be commercialized soon outside the United States.

Yet, the relatively recent domestication of forest trees has often met with public suspicion and even professional skepticism. Some people consider the aesthetic, recreational, and spiritual values of the forest more important than its value as a source of wood. Some of these values, particularly if virgin, primeval forest is the perceived ideal, do not fit well with any notion of domestication of forest trees. Yet forests continue to be a source of a variety of wood products, fuelwood predominating in some areas, and building materials and general industrial uses in other areas.

Social acceptance, both by the public and in a political context, must develop if we are to realize biotechnology's potential benefits. When ethical and religious concerns form the basis for some objections to genetic engineering, the positions taken can be widely divergent, vehemently held, and often perplexing to those with other beliefs or values. There can often be considerable scope for honest differences in both scientific and ethical views. It is also evident that these differences can be hardened by perceptions that commercial interests or political agendas are involved. That genetic engineering could be used to boost sales of a particular kind of herbicide or to force farmers to buy seeds from the controlling supplier feeds suspicion. Ironically, so-called terminator-gene technology, which would prevent farmers from using seeds from their current harvest for next year's planting by making transgenic crops sterile, could ensure genetic containment for many genetically engineered plants, including trees.

Several nongovernmental organizations aim to improve forest management practices by "green" certification schemes that are designed to influence consumer choice. Acceptance and approval of conventional tree breeding, clonal establishment, and the role of plantations in diverting exploitation pressures from natural forests have been slow and grudging by some certifying organizations. One organization, the Forest Stewardship Council, specifically prohibits transgenic trees in its program to certify well-managed forests and the resulting wood and paper products. Other certification schemes view the wood from transgenic trees the same as wood from trees produced by conventional breeding programs.

REGULATORY MECHANISMS

Whereas plants modified by traditional breeding practices and sexual techniques are unregulated, most countries do regulate genetically engineered plants produced by asexual approaches. Regulation typically is aimed at preventing a transgenic plant from being planted, used, transported, or disseminated until it has undergone a "deregulation" process that demonstrates it does not present unacceptable risks. After it is deregulated, a transgenic plant can be planted, harvested, processed, and traded just like any other plant.

The Biotechnology Regulatory Services program of the U.S. Department of Agriculture's Animal and Plant Health Inspection Service (APHIS) is the main agency that regulates the field testing, movement, and importation of genetically engineered organisms in the United States. APHIS issues various types of permits for each of these activities and evaluates petitions for deregulation to ensure that the plants do not pose a threat to U.S. agriculture or environmental health. To ensure compliance, field-test sites are inspected and records are audited. Joining APHIS in the regulation of transgenic plants in the United States are the Environmental Protection Agency for crops that have been genetically engineered for pesticide resistance, and the Food and Drug Administration for food crops.

Other countries are developing their own regulations to ensure both human and environmental safety, and it is not yet commonplace for countries to accept each other's protocols or deregulation processes, thereby allowing trade of a live transgenic plant or its seed. The Cartagena Protocol on Biosafety to the Convention on Biological Diversity, ratified by almost 100 nations, requires the labeling of live transgenic plants or seeds. In addition, live germplasm is also subject to the International Plant Protection Convention, which was created to prevent the introduction and spread of pests of plants and plant products.

The commodity produced by the transgenic plant (in this case, wood, pulp, or paper) is not generally subject to international trading restrictions unless it is considered a health or safety hazard. It is plausible, though, that the World Trade Organization (WTO) could become involved in disputes about trade of the products from transgenic trees, especially if the transgenic origin of the product is used to prevent free trade. WTO requires that any trade restrictions imposed to protect health and safety be based on science.

For public acceptance of genetic engineering, a trusted system of oversight and control seems essential. Examination and approval of professionally accepted practices for both the testing and the operational use of genetically engineered trees, covered by effective and workable regulations, will remain important. The most effective regulations will likely be those that provide for prudent risk management for transgenics by addressing all the elements of risk together. This will entail risk spread (for clones, transgenes, and transgene insertions), containment in the laboratory, containment and monitoring in test plantings, setting and policing conditions for commercial outplanting, evaluating performance of the plants, and monitoring for unexpected side effects. Regulatory decisions made on a case-by-case basis are likely to be most effective.

Chapter 7
Concluding Thoughts

Why should—or should not—biotechnology be applied to our future forests? Whatever the controversies over the technology used, two great, intertwined issues remain:

- how to meet the wood demands of an expanding and more prosperous world population without resorting to less environmentally benign materials; and
- how to preserve native forests and all their associated values.

Plantation forestry will clearly have a major role to play in addressing these issues, although it cannot in itself address the process of clearing forests for agriculture, except perhaps by offering an attractive economic alternative to such agriculture.

Only about 5 percent of the world's current closed-canopy forests are in plantations. The world's industrial wood harvest is about 36 billion cubic feet (1.6 billion cubic meters) per year, and an equal amount is harvested for nonindustrial uses, such as cooking and heating. About 35 percent of the former comes from plantations today, but the majority of the demand for nonindustrial needs is still being met by native forests. Although the total area of plantations is expected to rise to about 10 percent by 2050, even this potential supply is expected to fall well short of total wood demand. Thus, unless supplies from plantations increase well ahead of projections, the native forests will continue to be harvested, much of them in a nonsustainable way.

To achieve high yields, plantation managers must conduct intensive site preparation, apply supplemental fertilizers, control competing weeds, and manage stocking levels. In addition, the best regimes will use the best genetic stock available, including clones and trees engineered for optimum growth, form, adaptability to site, resistance to pests, and desired wood properties. The genetic technologies of conventional breeding, clonal deployment, and genetic engineering will almost exclusively be applied in plantations. The unintended consequences of these technologies are properly a concern, and good monitoring and good management are required to minimize them.

Meanwhile, it seems important to consider the needs of human populations that aspire to decent standards of living while seeking to enjoy natural forests and protecting the nonhuman members of forest ecosystems. The issues of ensuring a wood supply and preserving native forests can be addressed by reallocating substantial areas of forestland to three kinds of management. First would be minimally managed natural forests—wilderness, parks, and reserves with any human intervention designed to simulate natural ecological processes. Some wood harvest might occur in the reserves if it served conservation values, but wood production would not be a primary goal of reserve management. Second would be plantations focused on wood production. Here, efficient wood production would be the primary goal, although such values as wildlife habitat, high-quality water and air, and recreation would also be served. Third would be forests that serve multiple-use benefits for society along a potential continuum between minimal and intensive management.

Biotechnology, inclusive of genetic engineering, like any new technology, has its inherent risks and concerns. To the extent that the risks can be minimized, and insofar as it can enhance the role of plantations and help preserve natural environments, biotechnology will remain a viable management tool to address human needs.

Acknowledgments

For their valuable and valued help in preparing this booklet, we acknowledge with thanks Sally Atwater's enormous editorial contributions and the suggestions contributed by Steve Anderson, Dave Canavera, Julia Charity, Tom Conkle, Dave Harry, Jim Harding, John Helms, Steve Hill, Bob Kellison, Minghe Li, Iris Libby, Bob Luck, Rex McCullough, Steve McKeand, Tim Mullin, Chris Nelson, Pauline Newman, Robert Reid, Marc Rust, Nathan Scott, Steve Strauss, Armin Wagner, Christian Walter, Bob Weir, Phillip Wilcox, Ron Woessner, and Bruce Zobel.

Glossary

The following definitions of possibly unfamiliar terms were compiled and crafted from numerous sources. They include terms commonly used in forest genetics, and some definitions have been modified to make them both compatible with their usage in this booklet and understandable to a general audience. The page numbers indicate where in the text the term was first used and/or used in a contextually relevant manner.

allele an alternative form of a *gene* (at a given *locus*) differing in *DNA* sequence, 13
amino acid a building block of proteins, 48
angiosperm a true flowering plant that produces seeds enclosed in an ovary; the category includes *hardwoods* but not *softwoods*, 10
backcross to cross a *hybrid* to either one of its parental types, 19
base a molecular building block of *DNA*; four such bases constitute the four-letter genetic "alphabet", 49
breed a set of individuals within a species selected for a particular set of *characteristics*, 9
characteristic a specific expression of a *trait* (e.g., growth rate or disease resistance), 9
chromosome a microscopic, generally threadlike or rodlike body containing double strands of deoxyribonucleic acid (*DNA*), basically containing a linear sequence of genes, 12
clone a set of genetically identical individuals, produced by vegetative propagation, 3
dioecious having the male and female flowers (or *strobili*) produced on separate plants, 10
DNA deoxyribonucleic acid, a double-stranded, self-replicating acid of large molecular weight that is the genetically active portion of the *chromosome*, 47
domestication husbandry and guided *evolution* that suits human purposes, 47
embryogenic tissue a group of cells, some of which are competent to form an embryo, from which a plant can develop, 40
encapsulate to enclose a somatic embryo in a protective coating to produce an artificial seed, 40
evolution the change in the genetic makeup of populations of a species over time, usually applied to natural processes, 5
exotic a plant or species introduced from another country or geographic region outside its natural range, 5

extirpation the extinction of a local population of a species, 62
family a group of individuals directly related by descent from at least one common ancestor, 17
gene the smallest unit of genetic material governing the production and composition of a protein, 12
genetic engineering achieving new genetic combinations by means other than sexual reproduction, usually by *transformation*, 9, 47
genotype the genetic constitution of an organism, which together with the environment produces the *phenotype*, 29
gymnosperm a seed-producing plant whose ovules are naked instead of enclosed in ovaries, usually a conifer or *softwood*; gymnosperms produce *strobili* rather than true flowers, 10
hardwood an *angiosperm* tree, usually broadleaved, not necessarily with wood that is hard, 54
hybrid the progeny of genetically different parents; the term may be applied to the progeny from matings either between populations within species (*intraspecific*) or between species (*interspecific*), 19
inbreeding the production of offspring by mating related individuals, 4, 10
insertion the integration of short *DNA* pieces into intact *DNA*, 47
interplanting establishing young trees among existing forest growth, 3
interspecific between species, 19
intraspecific within a species, 19
line a succession of descendants from an individual or group of individuals, 53
locus (plural, **loci**) the physical location of the gene within the *DNA* strand in a particular *chromosome*, 12
migration the movement of *alleles* (*genes*) from one population to another, 13
monoecious having the male and female flowers (or *strobili*) in separate places on the same plant, 9
mutation a discrete change in the genetic constitution of an organism, often recognized as a sudden deviation from the ancestral *phenotype*, 13
outcrossing the intermating of unrelated individuals, producing outcrosses,10
ovule a structure that develops into a seed after its egg cell is fertilized by a pollen grain, 12
pathogen a parasitic organism directly capable of causing disease, 6
perfect flower a flower containing both pollen- and seed-producing organs, 12
pest any organism that damages a crop; pests can be insects or other animals, disease fungi, bacteria, viruses, or other plants, 6

phenotype the physical appearance of an organism as a result of genetic and environmental effects, 29

plantation a stand of trees established by planting or artificial seeding, 4

plus tree a tree selected on the basis of its outstanding *phenotype* but not yet *progeny tested* to determine its genetic worth, 23

progeny test a planting designed to evaluate parents by comparing the performance of their offspring, 23

propagule (1) a plant part, such as a bud, tuber, root, shoot, or spore, used to propagate a tree; (2) any individual resulting from either seed or vegetative propagation, 37

provenance (1) the geographic area and environment to which a population is naturally adapted; (2) the population concerned, 7

recombination the formation in the progeny of genetic combinations not present in either parent, 53

rotation the period between tree establishment and final harvest, 30

seed orchard a plantation consisting of clones or seedlings from selected trees for early and abundant production of seed, 23

selection (1) the process that favors the preferential survival and perpetuation of certain individuals over others; (2) a plant that displays one or more desirable characteristics and is selected for a specific use, 3, 15

self the offspring resulting from self-pollination (i.e., the pollination of a flower or strobilus with pollen from the same tree or clone), 10

softwood a conifer, almost always with needlelike leaves, not necessarily with wood that is soft, 71

somaclonal variation the higher-than-average rates of gene mutation that can occur during culturing of somatic cells, 44

somatic cell a body cell that is neither egg nor pollen (sperm in animals) or their precursor tissues, 40

strobilus (plural, **strobili**) the so-called flower of a *gymnosperm*, or conifer, 12

tissue culture the aseptic laboratory growth of cells, tissues, or organs, 39

trait a property or characteristic of an organism that may vary continuously (e.g. height) or discontinuously (e.g. alive or dead), 12

transformation the nonsexual acquisition of *DNA* by specific *insertions*; the customary form of *genetic engineering*, 50

transgene an *inserted DNA* sequence, 50

Suggested Reading

Ahuja, M.R., and W.J. Libby (eds.). 1993. *Clonal forestry. I. Genetics and biotechnology. II. Conservation and application*. Heidelberg: Springer-Verlag.

Burdon, R.D. 2003. Genetically modified forest trees. *International Forestry Review* 5(1): 58–64.

Burley, J., J. Evans, and J.A. Youngquist. 2004. *Encyclopedia of forest sciences*. 4 volumes. Oxford: Elsevier Academic Press.

Committee on Managing Global Genetic Resources. 1991. *Managing global genetic resources: Forest trees*. Washington, DC: National Academy Press.

Darwin, C. 1859. *On the origin of species*. Various reprinted editions.

Diamond, J. 1999. *Guns, germs, and steel*. New York: W.W. Norton.

Evelyn, J. 1664. *Sylva: or a discourse on forest trees and the propagation of timber in His Majestie's dominions*. London: Arthur Doubleday.

FAO. 2001. *State of the world's forests: 2001*. Rome: Food and Agriculture Organization of the United Nations.

Floyd, D.W. 2002. *Forest sustainability: The history, the challenge, the promise*. Durham, NC: Forest History Society.

Harper, J.L. 1977. *Population biology of plants*. New York: Academic Press.

Hunter, M.L. Jr., and A. Calhoun. 1996. A triad approach to land use allocation. In R.C. Szaro and D.W. Johnston (eds.), *Biodiversity in managed landscapes*. New York: Oxford University Press, 477–91.

Kumar, S., and M. Fladung (eds.) 2004. *Molecular genetics and breeding of forest trees*. Binghamton, NY: Food Products Press of Haworth Press.

Laarman, J.G., and R.A. Sedjo. 1992. *Global forests: Issues for six billion people*. New York: McGraw-Hill.

Larsen, C.S. 1956. *Genetics in silviculture*. London: Oliver and Boyd.

Li, M., and G.A. Ritchie. 1999. Eight hundred years of clonal forestry in China. I. Traditional afforestation with Chinese fir. II. Mass production of rooted cuttings of Chinese fir. *New Forests* 18: 131–59.

Libby, W.J. 1982. What is a safe number of clones per plantation? In H.M. Heybroek, B.R. Stephen, and K. von Weissenberg (eds.), *Resistance to diseases and pests in forest trees*. Wageningen: Purdoc, 342–60.

Lerner, I.M., and W.J. Libby. 1976. *Heredity, evolution and society*. San Francisco: W.H. Freeman.

Matthew, P. 1831. *On naval timber and arboriculture*. Edinburgh: Adam Black.

Mátyás, C. (ed.) 1999. *Forest genetics and sustainability*. Dordrecht: Kluwer.

Maunder, C., W. Shaw, and R. Pierce, 2005. Indigenous biodiversity and land use—what do exotic plantation forests contribute? *NZ Journal of Forestry* 49(4):20–26.

McCord, S., and K. Gartland. (eds.) 2003. *Forest biotechnology in Europe: The challenge, the promise, the future*. Research Triangle Park, NC: Institute of Forest Biotechnology.

McCord, S., and R. Kellison (eds.). 2005. *New century, new trees: Biotechnology for forestry in North America*. Raleigh, NC: Institute of Forest Biotechnology.

McKeand, S.E., T.J. Mullin, T.D. Byram, and T.L. White. 2003. Deployment of genetically improved loblolly and slash pines in the southern US. *Journal of Forestry* 101(3): 32–37.

Miller, J., S. Engelberg, and W. Broad, 2002. *Germs: Biological weapons and America's secret war*. New York: Simon & Schuster Touchstone.

Morgenstern, K. 1996. *Geographic variation in forest trees*. Vancouver: University of British Columbia Press.

Namkoong, Gene. 2005. *The misunderstood forest*. Vancouver: Forest Sciences Department, University of British Columbia. Available at genetics.forestry.ubc.ca/cfgc/proj_other/The_Misunderstood_Forest.pdf.

Renner, M. 2002. *The anatomy of resource wars*. Danvers, MA: Worldwatch Paper 162.

Sedjo, R.A. 2005. Global agreements and US forestry: Genetically modified trees. *Journal of Forestry* 103(3): 109–13.

Strauss, S.H., and H.D. Bradshaw (eds.). 2004. *The bioengineered forest: Challenges for science and society*. Washington, DC: Resources for the Future Press.

Strauss, S.H., A.M. Brunner, V.B. Busov, Ma Caiping, and R. Meilan. 2004. Ten lessons from 15 years of transgenic *Populus* research. *Forestry* 77(5): 455–65.

Shen, X. (ed.). 1995. *Forest tree improvement in the Asia-Pacific region*. Beijing: China Forestry Publication House.

Walter, C., and M. Carson (eds). 2004. *Plantation forest biotechnology for the 21st century*. Trivandrum, Kerala, India: Research Signpost.

Werner, D. 2004. *Biological resources and migration*. Heidelberg: Springer-Verlag.

Wright, J.W. 1976. *Introduction to forest genetics*. New York: Academic Press.

Zobel, B., and J. Talbert. 1984. *Applied forest tree improvement*. New York: John Wiley.

Zobel, B.J., and J.R. Sprague. 1993. *A forestry revolution*. Durham, NC: Academic Press.

Zohary, D., and M. Hopf. 1988. *Domestication of plants in the Old World*. Oxford: Clarendon Press.

Common and Scientific Names of Trees

acacia (the genus) *Acacia* spp.
almond *Amygdalus communis*
American chestnut *Castanea dentata*
aspen *Populus tremula*
birch (the genus) *Betula spp.*
black poplar *Populus nigra*
Chinese-fir *Cunninghamia lanceolata*
Chinese chestnut *Castanea mollissima*
coast redwood *Sequoia sempervirens*
contorta pine *Pinus contorta*
Douglas-fir *Pseudotsuga menziesii*
eastern cottonwood *Populus deltoides*
eucalypts (the genus) *Eucalyptus* spp.
European beech *Fagus sylvatica*
flooded gum *Eucalyptus grandis*
Italian cypress *Cupressus sempervirens* var. *sempervirens*
loblolly pine *Pinus taeda*
maritime pine *Pinus pinaster*
Mediterranean cypress *Cupressus sempervirens*
Monterey pine *Pinus radiata*
Norway spruce *Picea abies*
oaks (the genus) *Quercus* spp.
ponderosa pine *Pinus ponderosa*
poplars (the genus) *Populus* spp.
redwood *see* coast redwood
river redgum *Eucalyptus camaldulensis*
Scots pine *Pinus sylvestris*
Sitka spruce *Picea sitchensis*
shortleaf pine *Pinus echinata*
slash pine *Pinus elliottii*
strawberry guava *Cidium cattleianum*
sugi *Cryptomeria japonica*
willow (the genus) *Salix* spp.
yellow-poplar *Liriodendron tulipifera*

About the Authors

Rowland D. Burdon has been a researcher at the New Zealand Forest Research Institute (now branded Scion), Rotorua, since 1964. He has a B.Sc. from the University of Canterbury in New Zealand, a forestry degree from Oxford, and a Ph.D. from the Welsh Plant Breeding Station. His work has focused on quantitative breeding methodology, tree-breeding strategies, and the quantitative genetic organization of Monterey pine. This latter has involved genetic conservation issues and most aspects of the species' biology. In recent years, he has additionally focused on evaluation and management of risks for genetically engineered forest trees. He has worked in New Zealand, Australia, Chile, the southeastern United States, and British Columbia, was a visiting scholar at North Carolina State University, and has contributed widely to international scientific meetings. In 1992, he was elected Fellow of the Royal Society of New Zealand. Following official retirement in 2000, he is continuing work on part-time contract.

William J. Libby was a professor at the University of California-Berkeley from 1961 until 1994, with joint appointments in the Genetics, Forestry, and Conservation of Natural Resources departments. He has a B.S. in forestry from the University of Michigan and an M.S. and Ph.D. in tree physiology and population genetics from Berkeley, and he spent a postdoctoral year in the Genetics and Forestry departments at North Carolina State University. He has worked in forests in most states of the United States, and in Canada, Mexico, Venezuela, China, Australia, Fiji, and in many countries in Europe. During 1989–2002, he spent about half time in New Zealand, mostly at the Centre for Advanced Forest Biotechnology. He has taught courses at the University of Canterbury in New Zealand, the University of Washington, the Swedish Agricultural University, and the University of Zagreb in Croatia. He is now a senior associate with Zobel Forestry Associates of Raleigh, North Carolina.

Forest History Society Issues Series

The Issues Series published by the Forest History Society are booklets that bring a historical context to today's most pressing issues in forestry and natural resource management. FHS invites authors of demonstrated knowledge to examine an issue and synthesize its substantive literature. The Issues Series—like its Forest History Society sponsor—is non-advocacy. The series aims to present a balanced rendition of often contentious issues. They are attractive, informative, and easily accessible to the general reader.

AMERICAN FORESTS
A History of Resiliency and Recovery
by Douglas W. MacCleery

Many of today's forestry debates hinge on "how much there is" and "how much there was." *American Forests* documents the changes our nation's forests have experienced from colonial times to the present, covering Native Americans, population and agriculture, the industrial revolution, and early conservation efforts.

FOREST PHARMACY
Medicinal Plants in American Forests
by Steven Foster

The identification of taxol as a potential cancer fighting compound highlighted American interest in plant-derived medicines. Another fourteen plants are discussed in detail including ginseng, goldenseal, passionflower, mayapple, and bloodroot.

NEWSPRINT
Canadian Supply and American Demand
by Thomas R. Roach

Newsprint documents the growth of the Canadian newsprint industry and its traditional reliance on U.S. markets. The author discusses export restrictions and tariffs, government intervention, the changing structure of Canadian forests, international competitiveness and new approaches to fiber production.

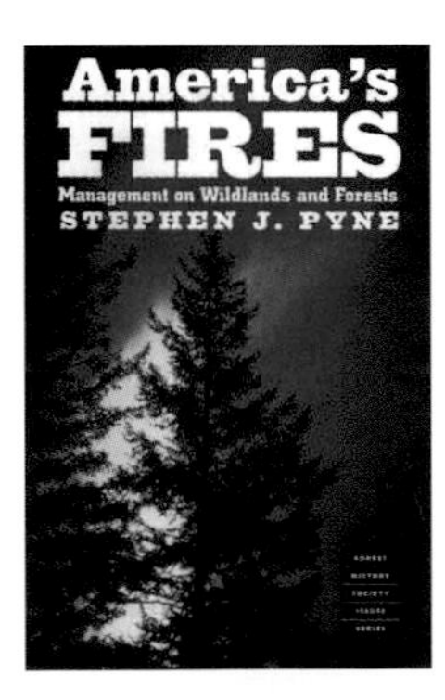

AMERICA'S FIRES
Management on Wildlands and Forests
by Stephen J. Pyne

Fire is at the heart of many forest health and sustainability issues being discussed today. Written by the authoritative expert on the subject, this booklet documents the extraordinarily successful twentieth century campaign to prevent and suppress wildland and forest fires.

FOREST SUSTAINABILITY
The History, the Challenge, the Promise
by Donald W. Floyd

This booklet suggests that forest sustainability on a global basis is a distant, worthy, and perhaps unobtainable goal without significant changes in technology and human population control. *Forest Sustainability* presents a historical context to the subject that helps the reader understand the nuances of this rapidly evolving topic.

CANADA'S FORESTS
A History
by Ken Drushka

A comprehensive overview of how humans have used Canada's forests in the past and the current state of those forests. It analyses changes in human attitudes toward forests and the author argues that, despite the centuries of use, the Canadian forest retains a good deal of its vitality and integrity.

Forest History Society Issues Series Order Form *Please send me:*

No. Copies	Title	Unit Cost	Total
________	American Forests	$8.95	________
________	Forest Pharmacy	$8.95	________
________	Newsprint	$8.95	________
________	America's Fires	$8.95	________
________	Forest Sustainability	$8.95	________
________	Canada's Forest	$8.95	________
________	Genetically Modified Forests	$8.95	________
Shipping & Handling ($3.00 for 1 book or $6.00 for 2–9)			________
		TOTAL	________

Call for discounts on orders of ten or more.

Name ______________________________

Address ______________________________

City/State/Zip ______________________________

Telephone (   ) ______________ E-mail ______________________________

☐ Enclosed is my check for $ __________ payable to the Forest History Society

☐ Charge $ __________ to my ☐ MasterCard ☐ VISA ☐ AmExp ☐ Discover

Card Number ______________________________ Expiration Date __________

Signature ______________________________

Please mail your check and this form to: Forest History Society, 701 Wm. Vickers Avenue, Durham, NC 27701
Or call: 919/682-9319 Order online: www.foresthistory.org

This form may be photocopied.